GENE HUNTER

WOMEN'S ADVENTURES IN SCIENCE

GENE HUNTER

the story of neuropsychologist

NANCY WEXLER

by Adele Glimm

Franklin Watts
A Division of Scholastic Inc.
New York • Toronto • London • Auckland • Sydney
Mexico City • New Delhi • Hong Kong
Danbury, Connecticut

Joseph Henry Press
Washington, D.C.

AUTHOR'S ACKNOWLEDGMENTS

To know you, Nancy Wexler, is to see passion for science in action, together with love for the human beings in need of that science. I am grateful to you for your inspiration and warmth and candor, for bringing your past to life so well that it seemed to be happening again as you spoke. I'm grateful as well to your patient and helpful colleagues Judy Lorimer, Julie Porter, Carl Johnson, and C. J. Li. I am deeply indebted to Alice Wexler and her book *Mapping Fate: A Memoir of Family, Risk, and Genetic Research*, which so clearly and gracefully maps the story of Huntington's disease, the generous Venezuelan villagers, and the stunning work of Nancy Wexler and the other Gene Hunters. Anne M. Pae spent hours introducing me to the Huntington's patients she cares for so lovingly, and I thank her. Finally, my thanks to the scientist I know best, my husband, Jim, for his challenging questions, his endless encouragement, and his belief that whatever the task, I can do it. —AG

Cover photo: Neuropsychologist Nancy Wexler removes a tissue sample from the freezer in her Columbia University lab.

Cover design: Michele de la Menardiere

Library of Congress Cataloging-in-Publication Data

Glimm, Adele.
    Gene hunter : the story of neuropsychologist Nancy Wexler / by Adele Glimm.
        p. cm. — (Women's adventures in science)
    Includes bibliographical references and index.
    ISBN 0-531-16778-X (lib. bdg.) ISBN 0-309-09558-1 (trade pbk.) ISBN 0-531-16953-7 (classroom pbk.)
 1. Wexler, Nancy—Juvenile literature. 2. Huntington's chorea—Genetic aspects—Juvenile literature. 3. Neuropsychology—United States—Juvenile literature. I. Title.

    RC394.H85G556 2005
    616.8'51042—dc22

                                                                        2005006645

NOTICE: No actual names of Venezuelan family members were used in this volume. Family members gave informed consent for the use of all photographs in which they were featured.

Any opinions, findings, conclusions, or recommendations expressed in this volume are those of the author and do not necessarily reflect the views of the National Academy of Sciences or its affiliated institutions.

Printed in the United States of America.
3 4 5 6 7 8 9 10 R 14 13 12 11 10

ABOUT THE SERIES

The stories in the *Women's Adventures in Science* series are about real women and the scientific careers they pursue so passionately. Some of these women knew at a very young age that they wanted to become scientists. Others realized it much later. Some of the scientists described in this series had to overcome major personal or societal obstacles on the way to establishing their careers. Others followed a simpler and more congenial path. Despite their very different backgrounds and life stories, these remarkable women all share one important belief: the work they do is important and it can make the world a better place.

Unlike many other biography series, *Women's Adventures in Science* chronicles the lives of contemporary, working scientists. Each of the women profiled in the series participated in her book's creation by sharing important details about her life, providing personal photographs to help illustrate the story, making family, friends, and colleagues available for interviews, and explaining her scientific specialty in ways that will inform and engage young readers.

This series would not have been possible without the generous assistance of Sara Lee Schupf and the National Academy of Sciences, an individual and an organization united in the belief that the pursuit of science is crucial to our understanding of how the world works and in the recognition that women must play a central role in all areas of science. They hope that *Women's Adventures in Science* will entertain and enlighten readers with stories of intellectually curious girls who became determined and innovative scientists dedicated to the quest for new knowledge. They also hope the stories will inspire young people with talent and energy to consider similar pursuits. The challenges of a scientific career are great but the rewards can be even greater.

Contents

Tracking a Deadly Disease

Nancy Wexler is a hunter. The big game she's spent much of her life pursuing is a particular gene, or unit of inheritance. People who inherit this gene from one of their parents develop a fatal illness called Huntington's disease. Nancy herself may have inherited this gene.

Nancy's long search for the Huntington's gene led her to villages in Venezuela where a greater percentage of people suffer from the disease than anywhere else in the world. There, year after year, Nancy and her team worked with large families—grandparents, parents, and children; uncles, aunts, and cousins. It's hard work, untangling family relationships, testing for symptoms, and working with the villagers whose blood and cooperation are critical tools for finding the cure. Nancy grieves when villagers become ill with Huntington's, because she knows how deeply they are suffering. She and they all share the same DNA and possibly the same mistake in that DNA—they literally are her family. But she never loses hope that science will someday cure the disease.

Nancy's work on Huntington's has taken her around the world, from London, England, to Papua, New Guinea. But ask her to name a great adventure and she'll tell you it's not exotic travel but rather science itself.

The search for the gene that causes Huntington's disease is a real-life detective story. As Nancy puts it, "There's a killer on the loose, and it's my job to find it before it claims more victims."

Huntington's is *always*
on her mind.

It may even be
in her *genes*.

THE DANCING DISEASE

I n a small village on the shore of a lake in Venezuela, it seems as if the whole population must be in the center of town today. Is it a special occasion, a festival of some kind? People are milling around near a low cinder-block building. Some are laughing and joking, some look more than a bit frightened, and everyone is sweating in the heat and humidity.

If you understand Spanish, you overhear comments like: "They say it will help us" and "They need so much blood!" and "I did it. You can do it." But listen closely and some of the shouted comments don't make sense at all; the words form meaningless phrases yelled over and over.

Near the doorway of the building, a woman with long, pale blond hair picks up a small child and hugs her. A thin young man hesitates to enter the building and the woman says gently: "Remember, I told you, of course you can still go fishing after Fidela takes your blood. No problem!" Then she hands some candy to a young girl exiting the building. "Good for you, Ana!" she says, nodding toward the small bandage on the girl's arm. "I'm proud of you."

The blond woman is Nancy Wexler, an American scientist, and right this minute she is hard at work on a research project that may eventually save thousands of lives. Maybe even her own.

Nancy and a Venezuelan child who's bravely contributed a blood sample jointly shout for joy *(opposite)*. At Lake Maracaibo in Venezuela, children as young as 6 years old *(above)* sometimes go fishing to help their families earn a living.

Many of the adults in this village are behaving rather strangely. Their arms and legs keep moving, even though they don't seem to be doing anything or going anywhere. Some seem to be dancing, though no one is watching their dance. These people are also unusually thin.

A few of the children move in unexpected ways, too. But rather than seeming to dance or jerk their limbs, these children move stiffly, arms and legs more like rigid boards.

Nancy emerges from the crowd, rounding the corner of the building, followed by children, her arms around the ones pressing close to her. At the side of the building, a young doctor is examining a short, husky man. The doctor isn't listening to the man's heart or peering at his throat. He's a neurologist and he's asking his patient to follow an upturned finger with his eyes and afterward to walk a straight line heel to toe. The man is having difficulty with these tasks. The doctor pats his shoulder and Nancy hugs him, saying, "Thank you, Luis. You're terrific!"

Luis walks away slowly, joining two teenagers, a girl and a boy, who are waiting for him. Nancy and the doctor confer. "Last year we weren't sure," the doctor says, "but now it's certain."

Nancy says, "And he has seven children."

"How are they doing?"

"The little ones seem O.K., so far." Nancy gestures at the young girl walking away with Luis. "But look how Maria walks."

The doctor nods. "A little stiff," he says sadly. "And when it comes from the father . . ."

"Yes," Nancy says. "Then it may hit early."

The doctor beckons to a woman waiting under a palm tree and Nancy moves away, still followed by children, many calling "Nancy! Look at me!" Again, Nancy stations herself near the doorway of the low building, watching people move in and out. A little boy who's been clinging to Nancy suddenly runs off and

Fidela Gomez, a nurse from Argentina, is so skilled at drawing blood and so amusing that the woman waiting her turn appears relaxed and at ease.

joins a woman who is moving wildly, throwing her arms around. The boy grabs a handful of her skirt and leads her slowly down a path between palm trees. Nancy thinks, *He's taking his mother home.* It is a normal thing here for children to take care of their parents, rather than the other way around, when their parents are some of the ones the villagers call *"perdido." Perdido* means "lost."

~ Hunting Down Huntington's Disease

What's going on here? Why are some of the villagers called "lost"? What's all this talk about blood?

Nancy Wexler is a kind of detective and she's working to hunt down a killer, a mass murderer, in fact. But her work has nothing to do with the police and the killer won't be captured at gunpoint. The killer is a defective gene, a unit of inheritance. It causes a disease called Huntington's disease. Huntington's, or HD, occurs in some 10 people in every 100,000. It has killed thousands of people all over the world. About 30,000 people in the United States are affected by HD. Some 150,000 more have a genetic risk of developing the disease. Singer-songwriter Woody Guthrie, the man who wrote the song *This Land Is Your Land,* is probably the most famous person to have had HD.

It's important to understand that HD is not a disease you can "catch," like chicken pox or the flu. It is an inherited or genetic disease and a type called an "autosomal dominant" disease. Autosomal means that both males and females get it. Dominant means only one parent has to have the disease for a child to have a chance of inheriting it. It also means that if the child inherits the disease, he or she will eventually get sick and die from it. (Some other genetic diseases, like cystic fibrosis, are "recessive," meaning the child needs to inherit abnormal genes from both parents in order to show symptoms.)

Each child in a family whose mother or father has HD has exactly the same one-in-two (or 50 percent) chance of inheriting the gene that causes the disease, no matter how many siblings are in the family. It's possible that none of the children of a parent with HD will inherit the gene; it's equally possible that all of them will. The phrase often used to describe this fact is "chance has no memory." (When you flip a quarter, the coin doesn't remember if it came up heads or tails on previous flips.) If a child of a parent with HD does not inherit the abnormal gene but rather the normal version of the gene, that person will never get sick with HD. His or her own children will never get the disease, because only the healthy gene is being passed down.

The symptoms of Huntington's usually begin to show up when people are in their 30s or 40s, but the disease can begin in childhood or even in old age. The first symptoms might be physical, such as twitching, stumbling, or making jerking motions. The uncontrollable movements are often described as "dancelike." When Nancy lectures on HD, she often explains the abnormal movements this way: "Seeing people with Huntington's is like watching a giant puppet show. Their limbs are jerked around as if by an unseen puppeteer, and there is nothing they can do about it." These movements only stop when the person is asleep. Other symptoms are mental and emotional; they may include memory problems, depression, and acting aggressively. Eventually, people have trouble swallowing, causing them to lose weight and waste away. There is no cure or treatment for Huntington's disease.

When the parent who transmits the HD gene to a child is the father, the disease sometimes shows up earlier than if the sick parent is the mother. There is a rare form of the disease in children, usually inherited from their fathers. That's why Nancy was concerned about Maria, the young daughter of Luis, who was walking stiffly. HD in children often causes rigidity or stiffness, rather than movements.

> "Seeing people with Huntington's is like watching a giant puppet show. Their limbs are jerked around as if by an unseen puppeteer, and there is nothing they can do about it."

As a scientist who studies Huntington's disease, Nancy is interested in its cause or causes, its symptoms, its distribution in the world, and how it might possibly be prevented, treated, and cured. One way she gains information is to collect blood samples from people in many different parts of the world. She takes blood

Posing inside their home, these family members from the village of Laguneta appear healthy, but some may be at risk for Huntington's disease.

samples from people who already have the disease, from their relatives who are at genetic risk, and from other relatives as well. She has worked in many places, including Naples, Italy; Barcelona, Spain; Shanghai, China; and Israel.

During the course of these scientific investigations, Nancy and the team she travels with have had many adventures. Often, each place they visit presents its own challenge. One day they flew in a tiny airplane to Papua, New Guinea, during a volcano alert. "It was very dramatic," she told her friend and colleague Judy Lorimer when she got back to her lab in New York. "All the trees had been cut down so as not to block escape roads. And there were wonderful— but scarily specific—signs posted everywhere telling what to do when the volcano blew its top."

In Palma, Majorca, where there were a large number of families with Huntington's, Nancy had expected to be able to use her basic Spanish. "The Majorcans spoke Catalan," she reported when she got back. "My Spanish didn't get very far. But luckily, we worked with a great scientific team from Palma."

And in tiny villages in the mountains of Peru, many miles outside Lima, the team kept getting flat tires with no replacements available. They visited families with HD who all lived in a string of villages along a main road but did not know of each other, even though they may be related.

~ The Worst and Best Place for Huntington's

Technicians in Boston analyze blood samples from Venezuela for hints on where to find the HD gene. The blood must reach the lab no later than three days after it is drawn.

Nancy has traveled to Venezuela more than to any other country. Every spring since 1979, she and a research team have flown there. These springtime trips are not about sightseeing and other holiday activities; rather, they are about families with Huntington's. Why choose Venezuela and, in particular, the little villages on Lake Maracaibo? Nancy and her colleagues found out that this region has more people with HD than any other place on Earth.

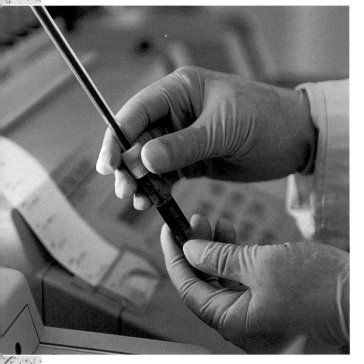

What Nancy wants more than anything in the world is to locate the gene that causes Huntington's disease. That gene holds the key to finding a cure for HD—which is what Nancy is after most of all! She hunts for this gene in the DNA found in the blood samples. She works with laboratories that analyze the blood, searching for clues that will tell scientists where the gene may be hiding. When the gene is finally identified, there will be a much greater chance for understanding the gene's mistake and how to cure it. The blood samples donated

6

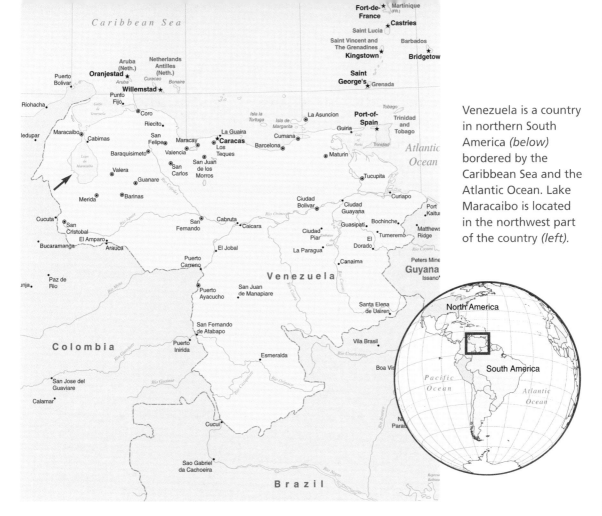

Venezuela is a country in northern South America *(below)* bordered by the Caribbean Sea and the Atlantic Ocean. Lake Maracaibo is located in the northwest part of the country *(left)*.

today by the people in the village will travel by plane tomorrow to a lab at Harvard University in Boston, where a scientist named Jim Gusella and his collaborators will analyze them, looking for clues that may lead them to the HD gene. The blood must reach the lab within 72 hours. Today was a "draw day," a good day for drawing blood because some members of the team are leaving tomorrow and can take the blood samples with them. It is impossible to ship them alone—and besides, they are much too valuable.

Late in the afternoon, Nancy relaxes in a simple outdoor café near the community center building. With her are Harvard University professors Anne Young and her husband Jack Penney who have spent the day examining, filming, and drawing blood samples from over 200 villagers. They drink Cokes. They've worked hard today as they have on every day of this year's trip. Are they any closer now to finding the gene than when the trip began, when all the trips here began? Nancy is certain they're getting closer. "I have to believe it," she says to Anne and Jack, "and I'm convinced it's true."

A child wanders over and sits on her lap. Another child chases a small pig down a path. "I have to believe it for the children," Nancy adds. "I look at them and want to believe they'll all keep developing and learning and have a future. But every year we come here and see that some of them are going downhill instead. Someday we'll celebrate finding the cure with all the families here."

Anne lifts her glass. She says: "To the gene! To the cure!" They drink to the gene and the cure coming quickly because they know they are in a race against time.

~ Working on Home Ground

It's a Wednesday morning and Nancy, instead of being off in Venezuela or another foreign country, is actually to be found in her office at Columbia University in New York City. Not that she's at her desk very much. She's mostly flying up and down the halls, her pale blond hair floating behind her, as she drops in to visit people in other offices—particularly her collaborators, Judy Lorimer and Julie Porter, plus scientists who work nearby. Judy is project director and Julie is data chief for the Venezuela project. Right now Judy and Julie are busy helping Nancy catch up with what she missed during a trip to England. Judy hands Nancy all the papers that have piled up while she was away. There are reports of important experiments, journals with articles on many hereditary diseases, invitations for Nancy to speak at conferences. Laughing, Nancy says to Judy, "There used to be a television program with the tag line 'Have gun, will travel.' Well, what I say is, 'Have science degree, will travel.'" Clutching an armful of work, Nancy gives each of her colleagues a hug and hurries back to her office.

Everyone Nancy encounters is at risk of receiving one of her hugs. She's known among her colleagues and friends for these big

> "We suddenly came upon two women, mother and daughter, both tall, thin, almost cadaverous, both bowing, twisting, grimacing."

hugs and for her brilliant smile, but most of all for her scientific work. Today Nancy's just back from London, where she visited the laboratory of scientist Dr. Gillian Bates, who has been studying some special mice that she developed. Gill genetically modified these mice to have symptoms resembling Huntington's disease, so they can be used to test new treatments as they are developed. When Nancy travels anywhere, just as when she's in Venezuela or "home" in her own office, she is constantly reading and writing about HD, discussing it with colleagues on the phone, in laboratories, or offices, lecturing about it, or visiting nursing homes where patients with the illness are cared for. Huntington's is always on her mind.

It may even be in her genes.

~ An Illness Like No Other

Huntington's disease led Nancy into the field of neuropsychology. You can figure out what a neuropsychologist does by taking apart the word. The prefix neuro comes from the Greek word *neuron,* which means "nerve." You probably know that psychologists study human behavior and the workings of the human mind. Neuropsychologists go one step further. They study the relationship between the central nervous system (the brain and spine) and behavior. They specialize in disorders of the brain that cause problems with thinking, emotions, and behavior. Huntington's disease is a disorder that fits this description perfectly.

George Huntington (1850–1916), son and grandson of doctors, was the first to tell the world about the illness named for him and that it is inherited from parent to child.

People always ask Nancy how Huntington's disease got its name. She tells them about George Huntington, the son and grandson of doctors, who lived in New York State during the 19th century. An observant child, he paid close attention to the inhabitants of his village. He particularly noticed some people who were behaving strangely. Later, after becoming a doctor himself, he wrote about his recollections: "We suddenly came upon two women, mother and daughter, both tall, thin, almost cadaverous, both bowing, twisting, grimacing. I stared in wonderment, almost in fear. What could it mean?"

George Huntington keenly observed the patients of his grandfather and father before him. Three generations of doctors in his family led George to be the first person in the world to describe accurately the hereditary pattern of Huntington's disease.

In 1872, when only 22, he published a paper on the disease, which has been known as Huntington's ever since. At first the illness was called "Huntington's chorea." *Chorea* is from the Greek word for dance, like the word "choreographer," someone who creates dances. You can understand why *chorea* was an appropriate word to describe this illness.

The evidence from Venezuela fills the filing cabinets in Nancy's office. The papers in these cabinets contain data on genes, blood, disease symptoms, and family relationships of thousands of Venezuelan villagers. Sometimes it seems as if, when a drawer is opened, the sound of many voices speaking Spanish fills the office.

There's other evidence of Nancy's Venezuelan trips as well. A photograph of a smiling little boy with brown skin and big blue-green eyes is propped on her desk. He is important, not only to Nancy but to thousands of people throughout the world. Later, we'll see why this is true.

~ A Scientist's Work

Today Nancy is wearing black pants, a bright orange knit top, and black flats. It's getting close to Halloween, which might explain the color scheme. The long amber beads of her necklace swing when she moves and so do her sparkly dangling earrings. Her blue eyes are as lively as her movements. When her collaborators have left, Nancy picks up the phone and taps in the number of her father across the country in Los Angeles, California. She tells him about the mice in the London lab, and they laugh together about the time she was raising white mice for an experiment when she was a kid. "How's your work going?" Nancy asks her father.

The Wexler sisters, Nancy *(left)* and Alice, have been jointly involved in the fight against HD for several decades. In 1995 Alice wrote *Mapping Fate,* a memoir about the Wexler family and HD.

His work? Milton Wexler is 97! Nevertheless, he still works part time, seeing psychotherapy patients.

In addition to psychotherapy, Nancy's father is also involved with Huntington's and they have many joint projects to discuss. He demonstrates another good thing about science: People retire if they want to, but they often don't want to. Many scientists become so interested in what they're doing they keep working as long as they can. In fact, scientists of any age have been known to say that their efforts often seem more like play than work.

Next, Nancy calls her sister, Alice, three years her senior. Alice also lives in California. She has a Ph.D. in history and has published a two-volume biography of Emma Goldman, a famous anarchist. Alice has also published a riveting memoir of the Wexler family and the origins of the Hereditary Disease Foundation. The book, *Mapping Fate: A Memoir of Family, Risk, and Genetic Research,* also describes the quest for the HD gene. She is now working on a new book, on the social history of HD. Today, Alice and Nancy discuss the researchers Nancy met with in London and the most recent advances in drugs that might treat the disease.

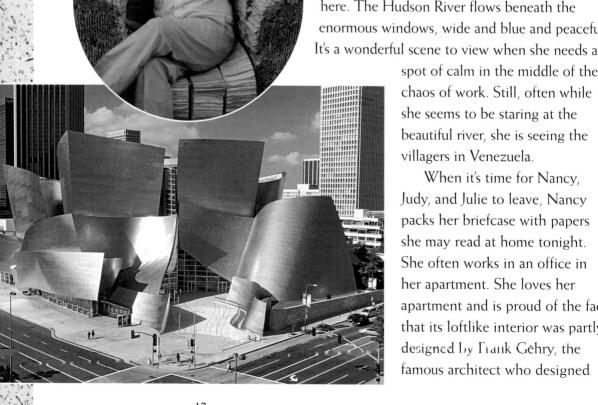

Nancy often spends time in Los Angeles. She's the president of the Hereditary Disease Foundation, a nonprofit foundation started by her father to find a cure for the disease after Alice and Nancy's mother was diagnosed with HD. The group raises funds for research on HD and related inherited diseases. The foundation also sponsors unique interdisciplinary workshops for scientists who work on HD and other genetic diseases. While she's in Los Angeles, Nancy stays at her father's apartment. She's a successful professional woman but still enjoys being a daughter living with a parent—at least some of the time.

Although world-famous architect Frank Gehry helped design Nancy's apartment, her home is not as eye-popping as Gehry's Walt Disney Concert Hall in Los Angeles!

But right now, here in her New York office, Nancy's life is almost too hectic. She's scheduled to give a guest lecture on genetics at the medical school at Columbia University. Later in the week, she'll lecture at the nursing school. At her desk, going over her notes for lectures, reading reports of new research, she raises her head now and then to stare out the window. The view from her corner office is so magnificent it's amazing that anyone can get any work done here. The Hudson River flows beneath the enormous windows, wide and blue and peaceful. It's a wonderful scene to view when she needs a spot of calm in the middle of the chaos of work. Still, often while she seems to be staring at the beautiful river, she is seeing the villagers in Venezuela.

When it's time for Nancy, Judy, and Julie to leave, Nancy packs her briefcase with papers she may read at home tonight. She often works in an office in her apartment. She loves her apartment and is proud of the fact that its loftlike interior was partly designed by Frank Gehry, the famous architect who designed the Walt Disney Concert Hall in Los Angeles and the Guggenheim Museum in Bilbao, Spain. She also mentions that Gehry is vice president and wife Berta Gehry is treasurer of the Hereditary Disease Foundation. Both are founders of the foundation and close friends of Nancy's.

But maybe she won't work tonight. She's been traveling so much and she's eager to spend some time with her long-time partner, Dr. Herbert Pardes. Herb is president and CEO of New York–Presbyterian Hospital, so he and Nancy have many medical interests in common. They also love to go to plays, movies, concerts, and the ballet when they can find the time. Maybe tonight they'll order Chinese food since neither of them cooks. After dinner Nancy will leave her papers unread and they'll go to the movies—one or two or three to make up for lost time. Nancy dashes down the hall to ask Judy and Julie to recommend some good movies. She's been traveling for too long a time to know what's happening in New York. She smiles to herself and thinks, *It feels really great to be home.*

Drs. Herb Pardes and Nancy Wexler believe that even the most dedicated scientists deserve an occasional vacation. Italy is one of their favorite places to relax.

Her parents probably never realized
just how useful *and* important

Nancy's stubbornness would be one day.

FAMILY SECRETS

For a long time, Nancy Wexler's childhood seemed close to perfect. She was born on July 19, 1945, in Washington, D.C., but the family soon moved to Topeka, Kansas. Life in Topeka included long summer days of swimming, plus winter fun sledding on inner tubes down snowy paths. Winter and summer, her mother read classic children's stories to Alice and Nancy before they were able to read themselves. A. A. Milne's tales of Winnie the Pooh and Christopher Robin were favorites.

Sometimes the family took cross-country trips to visit relatives in New York City. The girls looked forward to these trips. There were no cousins to play with, but her mother's three brothers—Jesse, Paul, and Seymour—were so entertaining that Alice and Nancy never missed other children. "My Uncle Seymour can do terrific tricks," Nancy told her friends. "He can spin coins around his fingers and make them come out his ears, his nose, or his pockets. He can make nickels stick to his forehead."

Uncle Paul played the saxophone. He had his own successful orchestra, featuring Uncle Seymour on clarinet. For the two Wexler girls, visiting the relatives was almost like having their own private circus.

Those trips to New York also offered opportunities to visit the fascinating exhibits at the American Museum of Natural History.

Like most young girls, Nancy enjoyed her dolls *(opposite)*, but life was more fun when big sister Alice was around *(above)*, especially on a good day for swimming at the lake.

Nancy's mother, Leonore, had taught biology and done scientific research before her children were born. To her an important part of parenting was opening her daughters' eyes and minds to the marvels of the natural world. Leonore loved watching her girls explore the same museum she herself had loved as a child, where enormous stuffed elephants gazed down over lifelike scenes of the African landscape.

Back home in Topeka, Nancy's mom shared her knowledge of biology by telling her girls the names of all the trees, flowers, and birds in their environment. Both Wexler parents welcomed pets of all kinds into the household. At any given time, Alice and Nancy might have had turtles, guinea pigs, parakeets, geese, rabbits, dogs, and many cats.

Leonore would often tell the girls about her days as a scientist in training at Columbia University in New York, where she had studied *Drosophila*, or fruit flies. *Drosophila*, she explained, are especially useful for research in genetics, the science of inherited characteristics. The flies reproduce so rapidly that many generations can be studied in a short time period.

The girls' father, Milton, was their second early instructor in science. He filled their heads with fascinating facts about physics

Most of the relatives in this photo are in Nancy's father's family, which is not affected by HD. Only her mother, Leonore *(second row, second from left),* and Nancy and Alice *(front row, second and third from left)* are at risk for HD.

and astronomy, explaining the motion of the planets with salt and pepper shakers on the dining room table. He was content to leave the biology to Leonore.

As the youngest member of the family, Nancy was stubbornly determined to grow up fast. At age four, she insisted on dressing herself in her thick, bulky snowsuit. Losing her balance as she tried to put her legs into the snowpants, she flopped straight onto the floor and got a bloody, scarred chin as a souvenir.

Nancy was equally stubborn at mealtimes. The Wexler family would sit at the dinner table and wait . . . wait . . . and wait some more while Nancy refused to eat or played with her food. She moved meat and vegetables in imaginative designs around her plate but rarely into her mouth. During these endless mealtimes, her parents probably never realized just how useful and important Nancy's stubbornness would be one day.

Just before Nancy turned six in 1951, the Wexlers sailed to Oslo, Norway, on the ship *Stavangerfjord.* There her father would spend the summer treating one of his patients. Milton Wexler had begun his adult working life as a practicing lawyer, but law didn't really interest him. So he went back to school and earned a Ph.D. in psychology. Eventually he became a psychoanalyst, a psychologist or psychiatrist who helps patients dig deep into their inner feelings and thoughts, with the goal of understanding their mental states, emotions, and actions.

The Wexler family enjoyed beach outings when Milton took time off from the work he loved. He obviously loved being with his wife, Leonore, and daughters Alice and Nancy as well.

In Topeka, Milton worked at the famous Menninger Clinic. There he treated combat veterans of World War II, taught other psychologists, and did research on treating people suffering from schizophrenia. He found this work very satisfying. The opportunity to take his family to Norway for a summer was a special benefit for all of them. Leonore recorded their experiences in her journal and in photographs that show a happy family. There was no hint of the trouble that would soon come.

~ Devastating News

The year before the trip Nancy's parents received information that shattered their world. For a long time the three uncles in New York—Jesse, Paul, and Seymour—had been feeling nervous, dropping things, and suffering from poor balance and memory problems.

Their hands and feet often moved uncontrollably. A neurologist, a medical doctor who treats diseases of the brain and nervous system, examined the brothers and diagnosed them as having an inherited disease he called simply "chorea."

Nancy's dad learned more about chorea from the doctors he worked with at the Menninger Clinic. They called it "Huntington's chorea" and reported that Nancy's mom, Leonore, was also at risk for the disease. Should she have it, Nancy and Alice might develop it as well.

Leonore recalled that her own father, Nancy's grandfather, had died of this disease when Leonore was only 15 years old. Since her father's family had come from Russia when her father was young, she really did not know who else in her family had the disease. When her father died, Leonore looked up Huntington's disease in the library and read "a fatal, inherited disease affecting only men." She panicked over the future fate of her three older brothers, whom she adored.

Even in 1950, Milton and Leonore heard from some doctors that only men could get the illness. Many prominent medical texts made the same claim— even though George Huntington himself had graphically described seeing a family of women affected with HD.

Uncle Seymour Sabin *(left)* could no longer play the saxophone after Huntington's began to affect the control of his physical movements. His brother Paul *(below)* became too ill with HD to continue leading his orchestra.

PAUL SABIN
and His Orchestra of 14 Pieces

GREENWICH COUNTRY CLUB
Saturday, August 29, 1936
Dinner and Dancing, 7:30 p.m. to 1 a.m.
$3.00 per person
Dancing only, 9 p.m. to 1 a.m.
$2.00 per person

Imagine the shock Milton and Leonore experienced at the realization that Leonore could become sick with a fatal illness—and that one or both of their daughters might eventually suffer the same fate!

For now there were the three uncles to worry about. "They'll need care and that will be expensive," Milton said. To earn the money his family would need, he left the Menninger Clinic and the Wexlers moved from Kansas to Los Angeles, where Milton could practice psychoanalysis and continue his research on schizophrenia. Too young at nine and six to face the brutal news, Alice and Nancy were told the family was heading west for a new adventure.

At first the Wexlers seemed happy in their new ranch house not far from the Pacific Ocean. Alice studied the piano (she has perfect pitch) and Nancy took ballet lessons. Their silver weimaraner, Sheba, was a loyal dog—even if one year she did eat the entire Thanksgiving dinner before the guests arrived! Through Milton's work, the family met an interesting group of people, including actors, musicians, writers, physicians, and scientists.

> The mother who had once identified every plant and bird they saw could only remember a few names.

The girls attended a special public elementary school run by the University of California for training teachers. The classes there were never boring. When the students studied the Pilgrims, they made Pilgrim clothes and re-created a Pilgrim school. When Nancy didn't learn the Lord's Prayer properly, she had to sit in the corner wearing a dunce cap with a pepper gag in her mouth—a Pilgrim punishment. The kids learned about African customs by building an African hut and cooking and eating African food. In sixth grade, Nancy represented Russia and China at a model United Nations.

But school lessons were not always easy for her. She had a tough time learning math and often says she had the biggest math block known to mankind. Once, as she struggled to multiply 13 by 13, she twisted her ponytail so hard that her teacher exclaimed, "Stop pulling your hair or it will fall off!" But dunce cap, math trouble and all, she always considered that elementary school one of the best educational experiences of her life.

~ Fleeting Knowledge

One day Uncle Seymour, the youngest brother, came to California for a visit. When he tried to do the magic tricks that had entranced Alice and Nancy in New York, the coins merely dropped to the floor. Uncle Seymour watched helplessly as his fingers danced and twitched. He seemed to trip over his feet. His speech was odd, too. When he played the clarinet, he sounded like a beginner. The girls could not understand what was happening to him.

Nancy and Alice also noticed that their mother seemed to be turning into a different person. She was afraid to make any decisions, even about new furniture. She grew more and more timid and unhappy. The mother who had once identified every plant and bird they saw could only remember a few names.

Nancy's mother, Leonore, could still enjoy a field of flowers, even when her scientist's knowledge of the natural world began to be eroded by HD.

Where had the knowledge gone? It was as if their mother was changing into someone else before their eyes. Had she been bewitched, like a fairy-tale princess? Her daughters didn't connect the change in Leonore with the change in Uncle Seymour, nor with the death of Jesse, the oldest uncle, or Paul, who died soon after.

~ A Young Researcher

As time passed, Nancy and Alice became less dependent on their mother's attentions. In their teens they tended to hang out at their friends' houses because Leonore was uncomfortable being with people who were not family members. Nancy and a group of her high school friends began spending a lot of time with a classmate named Michael Lorimer. (Today, Michael is married to Nancy's friend and co-worker Judy Lorimer.) Michael was a serious student of classical guitar and flamenco. He had been trained by Andres Segovia, one of the best guitarists in the world. Michael's fuzzy light brown hair had earned him the nickname "Tiger," and Tiger played a mean guitar. The kids loved to gather around him at school and in the backyard of his house. "Oh, the glorious hours we spent listening to Michael play!" Nancy remembered years later.

Nancy's high school was so large that there were 2,000 students in her tenth grade class. Like Los Angeles itself, the class was ethnically diverse. Later she transferred to another high school with smaller classes but less diversity, which was less interesting.

In a science class, Nancy read about an experiment she decided to duplicate at home. The idea behind the experiment was to see how touching and petting

Teen life in California was enlivened by "Tiger" Lorimer *(right)* and his guitar and by successful fishing expeditions, like the one Nancy proudly illustrates *(below)*.

can help people and animals handle stress better. The experiment measured the effect of touch and stress on mice and required her to get many baby white mice—which, fortunately, her parents did not object to. The idea was to ignore one group of mice, to pet a second group, and to create stress in a third group by banging on a can with the mice inside. The fourth group of mice would be stressed first, and then petted.

Nancy never got as far as writing up the results of the experiment. One day when her parents were out, she banged on the can. The mouse inside ran around, collapsed, and died in her hands! Nancy tried in vain to get its heart going. Her parents returned to find her sobbing because she thought she had killed the poor mouse. Her father stood over her, shaking his head. "You'll never be a scientist," he said. "It's hopeless." Nancy's eyes were swollen shut from crying.

Despite the mouse fiasco, Milton gave Nancy a part-time summer job helping him take care of some of his patients who had schizophrenia. Milton's treatment approach was to see his most disturbed patients daily so that he could provide regular and consistent input. He tried to keep people out of psychiatric hospitals as much as possible. He even bought a summer house on Lake Tahoe, where some of his patients spent the season, so he could treat them every day. Nancy helped her father take the patients out in a boat. One man had a habit of standing up in the boat and yelling swear words. A woman would shriek, "This is poison! This is poison!" whenever Nancy's mother offered food.

But it was not all work that summer. Nancy learned to water ski at Lake Tahoe. She met a boy who taught her how to catch and clean a fish. But the lessons took place at night in the boathouse. "No more of that!" Milton declared.

"And the winner is . . ." Nancy's prize high school essay on democracy didn't yet point to her scientific future, but it did reveal her belief in the dignity and importance of all people.

FIRST-PLACE ESSAY GETS A READING for appreciative audience as Nancy Wexler, winner of the 1960 essay contest sponsored by Pacific Palisades Junior Women's club, shares her ideas on "democracy." Deborah Bilsky, seated at right, wrote second-place entry. Winners are joined by their parents: Dr. and Mrs. Milton Wexler, left, and Mr. and Mrs. Sylvin Bilsky, right. In center, background: Mrs. Norman Loretz, contest chairman for Juniors and Mrs. H. Woodrow Linton, representing judges. Miss Wexler's essay will be entered by Juniors in district competition.

When high school started again that fall, Nancy attended some classes at the University of California at Los Angeles (UCLA). She took French and got the first D of her life.

~ The Deadly Destroyer

Even when the good days of their marriage and the marriage itself had ended, Milton continued caring for Leonore in every way through the difficult time to come.

In 1962, Nancy's parents decided to divorce. Neither of them realized that some of their difficulties might have been related to HD. Milton never remarried, and he and Leonore remained extremely close.

That summer Nancy flew to Mexico City with her father, where he attended a conference of psychoanalysts. At the conference, she listened to men and women from all over the world talk endlessly about their work. She watched them argue.

She felt their excitement as they shared ideas. She'd never before seen people so mesmerized by their work.

Swept up in this exciting environment, Nancy considered becoming a scientist like her father. A life combining love of work with interesting colleagues and usefulness to society was pretty appealing. She reminded herself that her mother, too, had once been a scientist—a fact that seemed at odds with the way her mother was now.

When it was time to apply to college, Nancy panicked and froze over the application forms. Her advisor said, "Just pretend you're writing me a letter." The tactic worked: Nancy was accepted at several first-rate colleges. She chose Radcliffe College, at that time the women's college of Harvard University in Cambridge, Massachusetts.

In 1963, Nancy graduated near the top of her high school class. A graduation day photo shows her in a sleeveless summer dress with a full skirt. Her hair is short and

wavy; she looks happy and excited. And why not? She was on her way out into the world, even though she had not yet tackled the question of what to study—nor what future to prepare for.

What will the future bring? What will I bring to it? Nancy considered these questions as her high school days drew to a close.

In the fall of Nancy's senior year, Paul, the middle uncle, died of Huntington's disease. The lively man who had once played banjo and led his own orchestra was dead at 58. Now, only one uncle, Seymour, was still alive.

These were the two strands of Nancy's life as she graduated from high school in her pretty summer dress in beautiful Southern California. First, there was the question of what to do with her life. Second, there was the disease that was slowly destroying her family. She never dreamed the two strands would soon be tightly intertwined.

Nancy loved
experiencing a different culture

and **talking**
to new people *about their lives.*

TAKING ON THE WORLD

A fter Nancy's high school graduation, she and Alice went to Guadalajara, Mexico, to study Spanish for the summer. Then it was time to set off for college. Nancy was concerned about moving so far away from her mother, who was trying to create a new life for herself after the divorce. But both parents urged her to go ahead and pursue her own new life.

This spacious house *(opposite)* is one example of the environment Nancy experienced in Jamaica. However, she preferred a home like the crowded but lively and friendly one pictured above.

~ Of Harleys and Lobsters

By choosing to study in Cambridge, a city next door to Boston on the East Coast, Nancy put the whole mainland United States between herself and home. Cambridge and Boston are both lively college towns, teeming with young people and activities. Nancy loved it all—her courses at Radcliffe, her friends, and the exciting life all around her. She was serious about her studies, but she made plenty of time for fun.

Nancy's boyfriend at the time was a good-looking Harvard student from Oklahoma who played in a rock band and rode a Harley around town. Nancy soon learned her way around the city, too—where to get the best (and cheapest) food, where to hear the best music, how to get to the Museum of Fine Arts or Symphony Hall.

Her life was full, but Nancy remained close to her parents and sister across the country. She wrote long letters, sending copies to each of them, and they kept in touch with her. Once, when she was failing a Serbo-Croatian literature course, she was grateful for her father's support. He said, "It's good for your character to fail once or even many times." He visited her in Cambridge, teasing Nancy that he had traveled there just for the seafood. "I can't help it if I love lobster," he joked.

At Radcliffe's Schlesinger Library, freshman Nancy had no trouble finding books on the wide variety of subjects that interested her, from folklore to psychology.

Because she had always been a last-minute person who completed her work just before the deadline, Nancy was panicked by the workload freshman year: So many papers, so little time! Her father rescued her once more by staying up all night typing a paper for her. Later he confessed that he probably should have made Nancy learn a lesson the hard way. But that was not Milton Wexler's style.

~ A Crucial Choice

All too soon it was time for Nancy to choose her college major. What subject should she focus on? And what career would it lead her to? Nancy had always been interested in her father's field—psychology and psychoanalysis. She remembered the charged energy that seemed to hover and crackle in the air above the analysts at that Mexico City conference in 1962.

After considering and rejecting several subjects, Nancy chose a double major: social relations and English. Social relations was equivalent to psychology at Harvard. The social relations major allowed her to get credit for many subjects she wanted to study: sociology, psychology, anthropology, and folklore. She wrote a paper on Brazilian Indian myths, considering them from the viewpoint of each of the subjects. She also took courses in philosophy, poetry, and architecture.

~ South American Interlude

While Nancy was just beginning college, her sister, Alice, had graduated from Stanford University in California. Alice won a Fulbright fellowship, which pays for recent college graduates to pursue their studies outside the United States. Alice chose to study social change in South America, at a university in Caracas, Venezuela.

At Christmastime during Nancy's freshman year—an especially colorful and festive holiday in Venezuela—Nancy went to visit Alice. Together the two sisters traveled to Margarita Island, where divers fished for oysters that might contain valuable pearls.

All in all, it was an exotic getaway. Nancy loved experiencing a different culture and talking to new people about their lives. She and Alice lay for hours on the beach and ate oysters under the stars. Who can blame them for thinking of Venezuela as a charming interlude, a kind of fairy-tale break from real life? Once jobs and families became the focus of their lives, the sisters believed at the time, it was unlikely they would ever return to Venezuela.

How could they know that Venezuela would play an important role in their futures?

~ Getting to Know Me

In her last year at Radcliffe, Nancy wrote her senior honors thesis on George Eliot—a female English novelist named Mary Ann Evans who wrote under a male alias in the early 1800s. Eliot's most famous novel, *Middlemarch*, traces the story of Dorothea Brooke, an intelligent young woman who struggles to find a life of meaning and purpose at a time when few options were open to women.

Nancy's advisor on her thesis was Erik Erikson, a famous psychoanalyst. He observed that Nancy may have chosen George Eliot because she secretly wanted to know herself better. Professor Erikson encouraged Nancy in her choice of topic: "Write about the tension and drama in Eliot's life cycle."

Nancy's thesis was a psychological study of how Eliot's books reflected her quest for independence in a culture that expected women to be submissive. Although Nancy Wexler was living in a time (the 1960s) when many new choices were opening up to women, she still had to make those choices. Studying George Eliot and her heroines shaped Nancy's thinking about her own future.

~ Living the Rasta Life

When she graduated from Radcliffe in May 1967, Nancy had no clue what she would do for a living. Working in some aspect of psychology or anthropology—the study of other cultures—seemed the most likely. Like Alice before her, Nancy was awarded a Fulbright fellowship to study in a foreign country. She chose to go to Kingston, Jamaica, in the West Indies. There she worked on a research project dealing with mental health. She attended sociology and medical school classes and worked with disturbed children at a clinic.

A market in Kingston, Jamaica, can be an exciting place to absorb the voices and customs of a culture new to a visiting young scholar.

Nancy was the only white person in the sociology class and frequently on the bus going to and from the university. This experience made her sensitive to how difficult it can be for people who are members of a minority group. At first she had trouble understanding the lilting variation of English spoken in Jamaica. She also had no friends, and that made her cry all too often.

When the people around her saw how lonely Nancy was, they befriended her. Soon she was being invited to parties and going out to hear reggae music. After sharing a house with a fellow Fulbright scholar, Nancy moved to a poor area of Kingston near where the Rastafarians lived. She shared the home of a Jamaican family, their six children, and a Peace Corps volunteer.

Her admiration grew daily for the family's generosity amid crushing poverty and its patience with her unending questions about life in Jamaica.

~ Hidden Dangers

In 1968, for the second half of her Fulbright fellowship, Nancy headed to London. There she studied with Anna Freud (daughter

of Sigmund Freud, the founder of psychoanalysis), who ran a clinic for the psychoanalysis of children. Anna Freud seemed to find the American girl with the long hair and short skirts kind of frivolous, but serious deliberations about the future were raging in Nancy's mind.

Nancy's term of study at the clinic left her wondering what form her future career would take. Would she be a psychologist helping patients cope with their lives? A psychoanalyst digging deep into the source of her patients' emotional states? Or a psychiatric anthropologist researching the mental makeup of entire cultures?

Psychiatrist Anna Freud (1895–1982) was a pioneer in the field of psychoanalysis of children. She taught Nancy about taking note of children's strengths as well as their frailties.

During her time in London, Nancy's father wrote her letter upon letter encouraging her to go to graduate school and prepare for a challenging career. Milton reminded his daughter how much happier he had become after trading a career that bored him—law—for one he loved with a passion. He knew that Nancy, too, would be happy only when she could use all her interests and talents.

In March of 1965 Nancy's Uncle Seymour died. He was the last of Leonore's brothers and the one she was closest to; his loss affected Leonore the most deeply. After Seymour's death, Nancy's mother was depressed and withdrawn. But Alice and Nancy, deeply involved in their studies, were still unaware that disaster would soon strike closer to home.

It seemed as if Leonore,
knowing her family history,

had perhaps understood
the truth all along.

"We Won't Give Up"

4

"Hey, lady!" the policeman yelled at Nancy Wexler's mother.
"Aren't you ashamed to be drunk this early in the day?"

It was 8 A.M. on a spring morning in 1968. Leonore Wexler had just parked her car and begun walking toward the Federal Building in Los Angeles. A minute ago she had been proudly en route to jury duty. Now she was paralyzed with horror.

Leonore drank very little—and certainly never in the morning. She knew she wasn't drunk. That meant she must be weaving, even staggering, as she walked.

Leonore had witnessed her three brothers suffering from Huntington's disease: They stumbled, tripped, and sometimes fell. That morning in 1968 she faced for the first time the terrifying possibility that she might have the ailment, too.

Milton Wexler arranged for Leonore to see a neurologist, who quickly diagnosed her with HD. Milton asked the neurologist to see Leonore over a number of sessions, to break it to her gently that something was wrong but not to tell her that the something was HD, nor that it was hereditary so that she would not worry about Alice and Nancy.

Leonore Wexler *(opposite)* had her own dreams, but her future was cut tragically short by the disease she inherited from her father and his family. Husband Milton Wexler *(above)* joined the effort with daughters Nancy and Alice to find a cure.

~ Breaking the Bad News

Soon afterward Milton phoned Nancy in London and Alice in Bloomington, Indiana. Would they come home to Los Angeles to celebrate his 60th birthday? Nancy found the request odd; her father had never made a big deal of birthdays. But she wanted to please him, so she came. Alice did, too.

At Milton's apartment the sisters learned the news that would change their lives forever. This was the moment Milton had been dreading since he first heard Leonore's diagnosis. After telling them the story of Leonore and the police officer who had thought she was drunk, he said, "Your mother has been diagnosed with Huntington's disease. Her brothers had it—so did her father—but her family always kept it a secret. Long ago she was told that only men could get it. Well, that's not true."

Shocked, Nancy asked, "What will happen?"

"It's a fatal disease," Milton told them sadly. "She may live for many more years, but eventually it will kill her."

"And I'm sorry to say, I have more bad news," he added gently. "Huntington's is hereditary, so you each have a one-in-two chance of having the same thing. Right now, there's no way of knowing what the future may bring."

**"We're not going to give up,"
Milton reassured his daughters.
"We're going to fight this."**

Dazed, the three Wexlers could do nothing but wrap their arms around one another for support. After a while Alice said, "A 50 percent chance to be healthy isn't so bad." Nancy agreed.

Then Milton told them the rest of the bad news. "If either of you does have it, there's a 50 percent chance you could pass it on to your children. Since there is no test, we can't predict the future for you or your children."

Nancy was too numb to think, but she decided one thing right away: *I will never have children unless I can be certain they are free of the disease. I will never risk passing the disease on to another generation.*

"We're not going to give up," Milton reassured his daughters. "We're going to fight this." With his ex-wife sick with a fatal disease and his two 20-something daughters at risk of meeting the same grim fate, Milton resolved to combat Huntington's disease with every fiber of his body. With great luck, a remedy might be found in time to help Leonore. If not, at least he could try to save Alice and Nancy.

This moment of sharing the news with his daughters changed Milton forever. Later in life, he explained his work on the disease this way: "I became a Huntington's disease activist because I was scared to death my wife and daughters would die of this ghastly disease."

~ Planning a Strategy

What was the best approach? Milton decided to find the most talented scientists in the fields of biology and genetic diseases and interest them in working on HD. He would supply them with the funds they needed to do intensive research.

At least two of Woody Guthrie's seven children inherited the abnormal HD gene. Some died young in accidents and may or may not have inherited the gene. Others are still apparently healthy.

The first step was to figure out how to raise that money. Milton joined forces with Marjorie Guthrie, the ex-wife of singer-songwriter Woody Guthrie. During his lifetime, Guthrie wrote about 1,000 folk songs. His work influenced the music of Bob Dylan, Bruce Springsteen, his own son Arlo Guthrie, and many others.

When Woody Guthrie died in 1967, he was only 55 years old. Guthrie had suffered from HD for about 15 years but was misdiagnosed as alcoholic or mentally ill. When the nature of his illness became clear, Marjorie decided to help search for a cure.

Woody had seven children genetically at risk for Huntington's, four of them with Marjorie. In 1967 Marjorie Guthrie, a former dancer with the Martha Graham company, founded the Committee to Combat Huntington's Disease (CCHD). The group would raise money for research and try to spark interest in the disease. In the fall of 1968, Milton started the California branch of the CCHD. He faced the same fears Marjorie did: Each of their children had a 50 percent chance of dying of HD.

Actors, singers, artists, and musicians who were friends of Milton's staged parties and concerts to raise money for research. In 1971 CCHD sponsored a "Tribute to Woody Guthrie" concert at the Hollywood Bowl. The concert was a sellout.

In 1968 Milton formed the organization that would become the Hereditary Disease Foundation. Its mission: to find treatments or cures for Huntington's disease and other inherited illnesses.

But Milton did much more than find the money for science. He found the scientists themselves and got them energized about fighting HD. He got senior and junior scientists in the fields of genetics and neuroscience to attend workshops where they swapped ideas and argued about the most important questions to study.

One of the best ideas to emerge from the first workshop was the notion that the search for a cure should involve young scientists. Open-minded and ready to brainstorm, these young men and women would be willing to consider every option and approach.

Before long, Milton Wexler's group broke away from Marjorie Guthrie's CCHD. While she was more interested in patient care, he was focused on understanding the ailment in hopes of finding a cure. In 1974 Milton formed the Hereditary Disease Foundation (HDF). Its mission: to find treatments or cures for Huntington's disease and other inherited illnesses. The foundation was a family affair. Both Nancy and Alice got fully involved in its activities. Both became trustees (members of the committee that ran the foundation).

~ The High Cost of Secrecy

In 1968 Nancy had begun graduate school at the University of Michigan. She intended to earn a Ph.D. degree in clinical psychology. Her education would include learning how to treat patients with emotional or mental problems, as well as conducting research in the field.

Nancy loved grad school, but her own emotions were in turmoil over her mother's illness. Leonore still seemed only slightly affected. But then Nancy would recall her three uncles and their untimely deaths and knew only too well what lay ahead.

About a year after her diagnosis, the family decided that Leonore should know the truth. It was Nancy who got the topic out in the air. While she and Alice and Leonore were all sitting around a friend's swimming pool, Nancy said, "I think it's great that Daddy's raising so much money for Huntington's research."

Although Nancy's mother's life was changing because of her HD, the mother-daughter closeness never changed.

"Yes," Leonore agreed, "that's a terrific thing for him to do. My brothers all had that disease—but not me."

"Well, that's what we all thought at first," Nancy said very softly. "But now it appears that maybe you do have HD."

Nancy held her breath. Yet her mother did not protest. It seemed as if Leonore, knowing her family history, had perhaps understood the truth all along. The entire family felt relief that they could now freely help Leonore and discuss the disease.

~ The Sunshine Solution

Family secrets, Alice and Nancy had learned, can make life harder for everyone. As she and Alice now realized, there had been way too much secrecy in her mother's family. Their three uncles had been called "nervous" rather than "ill," but in an environment of stigma and prejudice and fear and misunderstanding, perhaps this delicacy was understandable.

Nancy's University of Michigan studies drove home this insight. While training at a university psychology clinic, Nancy listened to a social worker describe the case of a troubled family. "The 10-year-old son is a disturbed child," the social worker said.

"Could there be a genetic problem in the family?" Nancy asked.

Annoyed by Nancy's suggestion, the social worker curtly dismissed it: "No, that's not it! It's all psychological." A year or so went by. One day the social worker came looking for Nancy. "Do you remember that 10-year-old boy in the troubled family we talked about?" she asked. "Well, his mother finally admitted that her own mother had died of Huntington's disease. She was terrified that she had it too—and that she had passed it along to her son. Once the boy's mother let the secret out, the whole family's life began to improve."

Nancy was gratified that the social worker had told her she'd been right about the boy's family. Also, hearing the story of that family reinforced the understanding Nancy had gained in her own family that shining "sunlight" on a problem is a powerful way to clear it up, even when there are no treatments.

For the time being, however, Nancy told none of her professors or fellow students at the University of Michigan that she might one day develop HD. Would they think she had no future? Maybe they wouldn't trust her to treat patients. *Come on, Nancy,* she said to herself eventually. *You know keeping secrets isn't the way to go; you have to start trusting people.* Gradually, she did tell her classmates and teachers. To her relief, everyone was understanding and supportive; she found herself feeling less depressed and helpless.

By the late 1960s, Nancy Wexler's work had begun to converge with her father's. Both Wexlers were intent on fighting Huntington's disease and other brain disorders. In 1969 father and daughter attended an international conference on psychoanalysis in Rome. A photo from the time shows Nancy and Milton smiling together among a crowd of conference attendees. The

In 1969 Nancy accompanied her father to the 26th International Psychoanalytic Congress in Rome. She never forgot the joy and excitement that scientists experienced in their work.

picture seems to illustrate their shared devotion to science and, to those who know their story, their special commitment to the science needed to combat the tragic condition savaging their family.

~ Encounter with a Slippery Brain

As Nancy's graduate studies zeroed in on HD, she became involved in setting up a "brain bank"—a collection of the brains of victims of the disease. Then she found herself actually examining the brains of patients who had died of the illness.

One day it was her job to wait for a cadaver, or dead body, that was being driven from Detroit to the University of Michigan in an ambulance and to collect the body's brain when the pathologist removed it. But the pathologist never showed up—only his assistant, called a deaner. He said, "I can take it out under your direction, Dr. Wexler." Nancy didn't dare admit she'd never seen a brain before, except in pictures. She'd never even seen a dead body before. So she just shrugged and gestured for him to go ahead. He peeled back the hair over the face. He was sawing away at the head, making a terrible noise. At last he separated the brain from its moorings and placed it carefully on the laboratory bench. "Would you like to take out the requisite tissues, Dr. Wexler?" he asked.

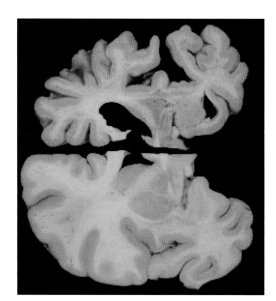

A cross section of the brain of a 48-year-old with HD *(top)* contrasts with a cross section of the brain of a 34-year-old without the disease. The destruction is obvious.

Nancy held her breath, panicking. What tissues? Where? She forced herself to speak calmly. "Oh, you go ahead and remove them," she said. "I'll put the tissues in the jars." The purpose of the research was to find out if HD was a slow virus. Nancy was supposed to process samples of the brain tissue, which required unscrewing the tops of jars of chemicals and adding the tissues. But she was wearing men's surgical gloves, which were much too big for her. She took off the gloves to better open the jars.

Suddenly, the brain started to slide off the table.

Instinctively, Nancy Wexler pushed the brain back into place with her bare hands. Then panic set in: *What if I've infected myself with the disease? What if Huntington's really is caused by a virus? That's what this experiment is all about!*

Nancy pulled herself together and managed to prepare the brain tissue samples to send to the lab of Nobel Prize–winner Dr. Carleton Gajdusek at the National Institutes of Health in Washington, D.C. There the samples would be inserted into chimpanzees to test the virus theory. Fortunately, the chimps stayed healthy! Nancy had been in no danger of contracting HD from that brain.

~ Meeting the Challenges of HD

Nancy decided to write her thesis, a major research paper required for the Ph.D. degree, on Huntington's disease. The paper focused on how it felt to be at risk for the disease. She did many in-depth interviews with people at risk as well as people with symptoms of HD and normal people as so-called controls to compare with the people at risk. She also administered tests of the cognitive, or thinking, abilities of people in all these groups. Some of her friends asked, "Don't you get depressed working on this?"

"No," replied Nancy. "I'm finding out how strong people can be, even with this possibility in their future. And I'm touched by the way people are willing to be open." The more people at risk or symptomatic for HD she talked with, the more determined she was to devote her working life to finding a cure. She was encouraged when her thesis project attracted the attention of many scientists, who then began to study the disease themselves.

As a Michigan grad student working at a university psychology clinic, Nancy had counseled patients seeking help with their problems. She was no longer a passive victim, waiting for other forces to decide her fate. She was really on the front lines, working to change her fate.

Each time Nancy returned home to Los Angeles, she could see that her mother was growing steadily worse. Leonore's fingers moved constantly. Nancy said, "It's as if she's playing a sad tune on a silent piano." Her toes twitched and jumped. Her left side sagged when she walked, and sometimes her legs couldn't hold her up. Worse, she was isolated, scared, lonely, depressed, and increasingly confused.

Nancy loved her mother deeply, and the older woman's relentless decline strengthened Nancy's decision of how to spend her own life. From now on, her private life and her professional life would be devoted to a single cause: stamping out Huntington's disease.

Within just *a few years,*
the **Lake Maracaibo** region

and its brave **inhabitants** would play
a major role in her life.

RISK AND DEATH

5

I n 1972 Nancy attended a conference in Columbus, Ohio, marking the 100th anniversary of George Huntington's original paper on the disease. Students of Dr. Americo Negrette, a Venezuelan physician, showed a brief black-and-white film he had made about a few small villages in an area along the shores of Lake Maracaibo in Venezuela. Large numbers of people in the region suffered from Huntington's disease. Indeed, no other place on the planet, it was believed, had so many people suffering with HD. Dr. Negrette, at work in this community, had noticed the symptoms and diagnosed the villagers with Huntington's disease.

Nancy embraced Lake Maracaibo villagers with Huntington's both literally *(left)* and through her dedication to research planned at workshops *(above)*.

Nancy never forgot the sad scenes of people with HD moving uncontrollably, of children clinging to obviously ill parents. Within just a few years, the Lake Maracaibo region and its brave inhabitants would play a major role in her life.

~ Creative Energy

It was 1974 and Nancy's student days were over at long last. Dr. Nancy Wexler, with her newly minted Ph.D. in clinical psychology, headed to New York City. She had landed a job as an assistant professor of psychology at the New School for

Social Research. (Today the college is called the New School University.)

The New School was located in the Greenwich Village neighborhood of Manhattan. For decades the village had been famous as the home of writers, artists, and other creative, free-spirited people. It was a great neighborhood for eating foods from many cultures, listening to jazz, and shopping for crafts and one-of-a-kind clothes. Nancy loved the area and especially enjoyed the company of her artist friends. The lively street life was a welcome contrast to L.A.'s car-oriented culture. In Greenwich Village, Nancy could walk everywhere or hop a subway to other parts of Manhattan.

Washington Square in New York's Greenwich Village is a magnet for young people and the not-so-young alike, a place that represents freedom and creativity to all.

Nancy liked her new job teaching students, many of whom were adults returning to school. But she also had another job. She had assigned it to herself, and it paid no salary. She traveled to workshops of the Hereditary Disease Foundation (HDF) wherever they were held, seeking out researchers who attended the meetings. Just meeting this vibrant young woman and knowing that she was at risk of developing HD inspired many scientists to find out more about the disease.

HDF workshops were run differently from those at most scientific conferences. Rather than presenting ideas that had been thought out carefully or tested in a lab, attendees poured out any and all ideas that occurred to them: Impossible notions. Mind-boggling brainstorms. Off-the-wall suggestions that would never fly in a rational world.

"Just let yourselves go," Nancy told the researchers who attended these workshops. "Make free associations. We want to hear your wildest ideas, so don't be afraid to make a mistake. After all, at this stage, who can tell a mistake from a useful idea?"

Milton Wexler had pioneered this innovative approach. Many of his friends and patients were artists and other creative types, and Milton admired the way they tackled their work. Why shouldn't science be approached in the same way that books got written, movies got made, buildings got built, or music got composed? The younger scientists in particular embraced this strategy.

In the early HDF workshops as in real life, not all fresh, creative ideas led to successful results. Many workshop participants pursued scientific roads that culminated in dead ends. Despite the lack of concrete results, Milton's many theater and artist friends seemed happy to help the work of the foundation. When some of them gave parties for the scientists, the researchers went home talking about their new acquaintances, such as actresses Carol Burnett and Julie Andrews and architect Frank Gehry.

Almost all the workshops began by introducing a family with HD to the scientists. Researchers learned firsthand how devastating the disease is—affecting all aspects of body and mind and getting worse every year. They could see how all members of the family suffer from the reverberations of this tiny mistake in the gene. Researchers developed a passion for finding a cure as fast as possible!

At workshops on HD, ideas on promising routes for research flew among the scientists. This workshop was supported by actress Jennifer Jones Simon *(second from left)*.

~ A New Approach: Hunting for a Marker

So far scientists had discovered that HD slowly destroys the basal ganglia. These are large groups of brain cells in the middle of the brain that control movement. The disease also damages the part of the brain called the cortex. Damage to brain cells in the cortex produces mental problems and difficulties with speech and memory.

The basic question crying out for an answer was: How and why do these brain cells die? David Housman, a molecular geneticist at the Massachusetts Institute of Technology in Cambridge, thought the best way to find the answer was to identify the abnormal gene responsible for Huntington's disease. Nancy agreed. "We need to find the root cause of this devastation and stop it in its tracks. Like going to the source of the Nile to fix a problem—rather than trying to stop all the flooding downstream."

> A marker would provide a hint about where the HD gene might be in the DNA.

Before long it became clear that the best way to study HD was to search for a "marker" that would provide a clue about the defective gene's location. What is a marker? Why would finding one be helpful? A marker would provide a hint about where the HD gene might be in the DNA.

What does a marker do? A marker narrows the territory. It tells you, "Look in this continent, this country, maybe—with luck— even this city." But finding markers is hard to do also.

Nancy describes the situation like this: "Imagine an earthquake at the South Pole where many penguins are sitting on an enormous ice floe. When the ice floe breaks up, one pair of penguins who were already far apart will likely end up floating on separate pieces of ice. But two penguins hanging out next to each other will probably end up on the same piece of ice. If one of these penguins is a DNA marker and the other is the HD gene, and they are on the same floe, the two will travel together. So finding a DNA marker will tell you something about where you might find the gene."

What Are Genes and Where Are They, Anyway?

What exactly is a gene? Genes are made of DNA, which stands for deoxyribonucleic acid. DNA is a very long molecule, shaped like a spiral ladder, called a double helix. A gene is a long string of DNA. Each person has 3 billion rungs on the DNA ladder. Only a tiny fraction of our DNA, less than 1 percent, is packaged into the 25,000 or so genes that run the show—they control all of our bodily functioning as well as how we look and our susceptibility to disease. Our genes have specific addresses on our 23 pairs of chromosomes. We all have two copies of each chromosome: one from our mother and one from our father. Our chromosomes are located in the nucleus of almost all of our cells. The nucleus is the cell's control center.

Sometimes a mistake causing a genetic disease is very tiny. It could be that one rung on the DNA ladder is in backwards, too big, missing, or changed in some way. If all of the DNA in our body has 3 billion rungs, then looking for a single damaged gene is like looking for a lone killer in half of the world's population of 6.5 billion people. And you don't even know where to start!

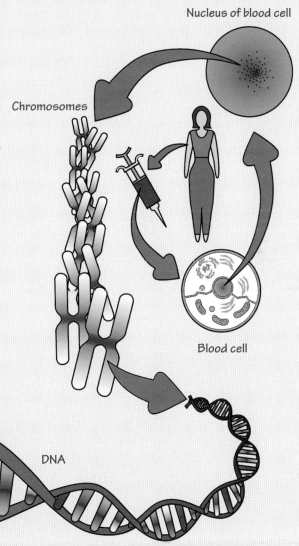

Nucleus of blood cell

Chromosomes

Blood cell

DNA

Scientists searching for the gene that causes HD—with no idea of where it might be—is like the police trying to find a criminal somewhere on the planet without a clue as to where to look. Or as Nancy said: "It's not as easy as finding a needle in a haystack because at least you prick your finger when you find a needle. It's more like finding a certain piece of hay in that haystack."

Nancy and the other scientists working on HD needed to find markers that were closely spaced throughout all of our chromosomes. If this were the case, the markers might zero in on the gene the same way the police could zero in on a criminal. If, for example, they knew the suspect was in Maple Grove, New Jersey, then instead of looking all over the United States, they could hunt door to door in that town.

But the real process would be even more complicated. To find out if a marker was close to a gene, researchers would have to see if the marker and the gene could be observed together in a family. People who had Huntington's would have one form of marker; their relatives who did not have the disease would have a different form of the same marker.

Although researchers had identified and studied a few American families with the disease, they had not located any large groups of big families with many cases of HD. As Nancy considered the problem, she recalled the film she had seen in 1972—the one showing a cluster of HD families who lived in the Lake Maracaibo region of Venezuela. Could she get an expedition funded and organized to go to Venezuela? How would the people there feel about being studied? These were big questions with uncertain answers.

~ Nancy Is the Boss

Meanwhile, Nancy was tackling yet another challenge. In 1975 the U.S. government formed a group called the Congressional Commission for the Control of Huntington's Disease and Its Consequences. The goal of the commission was to develop a full picture of HD in the United States: How many people had the disease? Where did they live? How was the disease affecting their lives?

In 1976 Nancy was asked to be the executive director of the commission. Although she worried she might not be up to the task, her father, sister, and fellow scientists encouraged her to take the position. Nancy left her teaching job in New York and moved to the city of her birth: Washington, D.C.

Like the Hereditary Disease Foundation, the Congressional Commission for the Control of Huntington's Disease was, in a way, a family affair for the Wexlers. Milton held the position of vice chairperson. "My own daughter has become my boss!" he liked to joke. Marjorie Guthrie, Woody Guthrie's ex-wife, was the commission's chairperson.

When Nancy testified before Congress, as head of the HD commission, she might have been the only scientist who understood the disease both personally and professionally.

Among its many other jobs, the commission collected the personal stories of those with HD. At public hearings, people told of their experiences. Doctors testified about their patients. Government health officials described cases in their districts.

The most poignant stories came from families whose lives had been affected by the sickness. "Sometimes people came to talk to us who thought they were the only HD family in the world," Nancy said to her sister, Alice. "At our hearings, they met other Huntington's families for the first time." Nancy marveled at the courage and strength of families coping with this extreme adversity.

The commission eventually produced a 10-volume report on HD in the United States. The study focused on the suffering the disease caused—and the lack of help available to treat it. Nancy took pride in the report. Finally, individuals and institutions throughout the country were paying attention to HD. More research money became available. More support was available for patients and their families. Most important of all, more scientists began to devote their time and talents to research focused on combating the illness.

In some ways, these scientists were risking their careers. Researching an inherited disease in human beings could mean years of effort with few or no results. Enormous genetic differences exist among people. Humans do not come in the purebred versions that lab animals do, so the results of studies involving people are less clear-cut. And Huntington's symptoms rarely emerge before middle age, making it hard to find several living generations of a single family to study.

~ Do Homozygotes Hold the Answer?

Nancy and other scientists wondered if finding homozygotes for the abnormal HD gene—people who had inherited the abnormal Huntington's disease gene from both their father and their mother, rather than from just one parent—would advance the research. Research into a different disease, called familial hypercholesterolemia, had shown that a child who inherits that illness from both parents has a far worse case of the disease. Familial hypercholesterolemia causes extremely high cholesterol levels—and, because of those, heart attacks. In one case, a daughter who had inherited the gene for the disease from both her parents suffered a heart attack at the age of seven! Nobel Prize winners Michael Brown and Joseph Goldstein studied this family and it helped them crack open the problem of heart disease. Could homozygotes help with Huntington's disease? Where were these families and patients to be found?

> Most important of all, more scientists began to devote their time and talents to research focused on combating the illness.

"We may be able to pinpoint the cause of Huntington's by studying homozygotes," Nancy suggested at a workshop. "There would be no normal gene to hide how the defective gene works."

Once again, Nancy recalled the film about Huntington's in Venezuela. Wasn't it likely that a place with such a high concentration of the disease would have residents who had two parents with HD? Other scientists, too, believed the communities

What Is a Homozygote?

If a child's parents *both* have HD, instead of just one parent, how does that affect the child's chances of inheriting the disease? To answer this question, it helps to understand the difference between a *homozygote* and a *heterozygote*.

Remember that every individual carries two sets of 23 chromosomes, one from each parent. The gene for Huntington's disease resides on chromosome 4. If a child inherits a defective copy of this gene, he or she will develop the disease.

Let's say that in a particular family both parents have Huntington's disease and that each parent carries one normal (unharmful) gene and one abnormal (harmful) gene. Because each parent carries one abnormal gene and Huntington's disease is a dominant disorder, the parents will both become ill with the disease and eventually die. Their children have four possible ways of inheriting their parents' genes *(See diagram below.)*. A child will have a 25 percent chance of inheriting the normal HD gene from the mother and the abnormal gene from the father; a 25 percent chance of inheriting the abnormal gene from the mother and the normal gene from the father; a 25 percent chance of inheriting the normal gene from both parents; and a 25 percent chance of inheriting the abnormal gene from both parents.

A child who inherits two of the same kind of gene—either the normal gene or the abnormal gene—is called a *homozygote*. A child who inherits two different kinds of gene—one normal and one abnormal—is called a *heterozygote*.

Don't forget, though, that it only takes one abnormal HD gene to cause the disease, so children of parents who each carry one abnormal gene have a 75 percent chance of getting the disease themselves. The only child who would escape the disease is one who inherits both normal copies of the HD gene. Unfortunately, the homozygote with two copies of the abnormal gene has a 100 percent chance of passing down the disease to his or her own child. There's no normal gene to pass on.

Would homozygotes for the abnormal HD gene show differences in the effects of the disease? Would they display different patterns of symptoms? Would they be younger when the sickness began? These are some of the questions researchers wanted to study when possible homozygotes with Huntington's were identified.

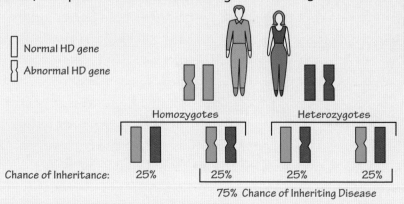

Normal HD *gene*
Abnormal HD *gene*

Homozygotes   Heterozygotes

Chance of Inheritance: 25%   25%   25%   25%

75% Chance of Inheriting Disease

around Lake Maracaibo might hold the key to the HD mystery. The congressional commission on Huntington's disease therefore recommended funding a Venezuela project.

The prospect of investigating the Lake Maracaibo region excited Nancy. Events in her own family now conspired to deepen the kinship she felt with these faraway strangers.

~ A Sad Ending

By the spring of 1978, Leonore Wexler's condition was much worse. "Uncontrollable motions racked her frame," Nancy said. "When she sat, her spasmodic body movements would propel her chair along the floor until it reached a wall; her head would then bang repeatedly against the wall. To keep her from hurting herself at night, her bed was padded with lambswool."

Nancy's mother continued to lose weight. Like most Huntington's patients, she needed to consume as much as 5,000 calories per day (the norm is 2,000). This may have been because she moved so much, or it might have had something to do with the unique metabolism (the way the body uses energy) of HD sufferers. Nancy, Alice, and Milton brought her bags of high-calorie foods—candy, ice cream, cookies—but Leonore continued to waste away. "My mother doesn't gain any weight," Nancy told a friend. "But I do. I eat to keep from crying."

One terrible day in early May that year, the family was informed that Leonore had died. Alice rushed to the nursing home where she had been residing—only to find her mother was still breathing. Failing to notice her very shallow breathing, the nursing home had mistakenly declared her dead.

After Leonore was gone, wonderful childhood memories, like this vacation in the West with their mother, comforted Alice and Nancy.

But on Mother's Day in 1978, Leonore Wexler finally passed away. At a private family service they held for her, Alice and Nancy took turns reading aloud from letters Leonore had written to friends when she was a healthy young girl. As they read, the sisters felt close to the happy, lively woman their mother had once been.

The Wexlers scattered Leonore's ashes in the Pacific Ocean, saying their final good-byes as her remains disappeared beneath the waves. Though their mother was gone, the family's real tribute to her had begun long before her death—and would endure long after. The battle against the brutally efficient killer that was Huntington's disease would continue to be waged in memory of Leonore Wexler.

One of *the team's first* **tasks**

was to construct

a **"*pedigree*"** *of the villagers.*

6

"WE ARE ALL ONE FAMILY"

I n 1979 Nancy journeyed to Venezuela, together with her
colleague Tom Chase. Tom was in charge of research at the
National Institute for Neurological Diseases and Stroke, part
of the National Institutes of Health. A scientist who believed in
encouraging women's careers, Tom was one of the people who'd
offered Nancy the job with the government commission. This
was Nancy's first trip to Venezuela since she'd visited as a carefree
college student. Back then she had enjoyed the festive customs of
a South American Christmas.

This time, however, "festive" was the last word Nancy would
have chosen to describe the world around her. Huntington's disease
seemed almost the norm in this little pocket of Venezuela near Lake
Maracaibo. Everywhere she looked, she saw bone-thin, staggering
people, often with limbs moving wildly. Many seemed mentally
troubled, occasionally addressing unseen, imaginary inner voices
and presences. Everyone, sick or healthy, lived in poverty. The days
were hot and steamy, with no air conditioning to provide relief.

Nancy watched a man with a lolling tongue make his way down
the road in the village of San Luis. His stick-thin arms flailed in the
humid air. His skinny, bent legs stumbled along, tracing a crooked
path. Nancy Wexler held her breath, afraid that the man might
topple over at any moment.

The pedigree chart
(opposite) showed
Nancy's team how the
Venezuelan villagers
were related to one
another and who
was affected or at risk
for HD. It eventually
included over 18,000
individuals. Would you
have guessed that this
porch *(above)* was an
office? From here
Nancy directed her
team's visit to Laguneta.

Dr. Thomas N. Chase of the National Institutes of Health believes in Nancy and believes in the importance of solving the mysteries of Huntington's disease.

It was one thing to watch a grainy black-and-white film of these men, women, and children. It was quite another thing to walk among them, touch them, and speak to them yet be completely unable to help them.

Two facts comforted Nancy. First, the people with the disease were living out their lives in the heart of their communities. They lived at home with their families, they sat in front of their homes, they wandered the narrow streets and were gently guided home again. The sick were not segregated and hidden away in nursing homes. Of course, there were no nursing homes here. But in some ways the whole village was a nursing home.

Second, Nancy felt close to her mother once more. She said to Tom Chase: "I've seen many Huntington's patients, but that woman there moves exactly the way my mother moved. I can't stop looking at her. It's like seeing my mother in front of my eyes."

Fortunately, Americo Negrette, the Venezuelan physician-scientist who first discovered and diagnosed HD in these communities, came with Tom and Nancy to make introductions. His two students, Ramon Avila-Giron and Ernesto Bonilla, also accompanied Tom and Nancy. At first, Avila-Giron and Bonilla were the ones who made the strange requests for information. Their participation made all the difference in the world.

Tom and Nancy, both blond and blue-eyed, could not have been more noticeable among the mostly tanned, dark-eyed inhabitants of the Venezuelan villages. Neither of them knew enough Spanish to explain clearly who they were or what they wanted and were grateful for help. But they knew what they were looking for: someone with Huntington's symptoms whose father and mother both had the disease—and both of whose parents were still alive. That person might have inherited the defective

Huntington's disease gene from both parents and, if so, would be a homozygote for the abnormal HD gene. If the parents were still alive, blood samples from both parents and their offspring might be studied to learn more about HD.

Through their colleagues, Tom and Nancy questioned the villagers as politely as they could. After all, they were complete strangers, asking for very personal information. Would people cooperate with them? Surprisingly, the villagers freely answered their questions. Disappointingly, they knew of no one in the village who met the requirements.

"Yes, there is a family like that," said one woman. "But the mother died in the spring."

"I know a boy who is ill and has a sick mother, too," said one man. "But his sick father moved away."

Nancy asked, "Who knows most about the people of the village?"

A woman answered, "There is an old man who knows all about everyone. You can find him in the shop." Nancy, Tom, and their group followed the woman's directions to the only shop. As they walked, they were surrounded by curious villagers. Many of them—even some small children—showed symptoms of Huntington's disease. The children's bodies did not move constantly like the bodies of the adults with HD, but rather moved stiffly.

Team members Jacqueline Bickham *(standing)* and Anne Young, along with Nancy, interact closely with Venezuelans because friendship is important as well as science.

When Nancy and Tom found the old man, he told them that his father had died of HD. Though he himself was at risk, he was still healthy. "I have 34 children," he said proudly.

"How many wives?" Tom asked.

"I have three wives. Twenty-five of my children are with two wives who are first cousins."

He went on to tell them where they could find the family they were seeking. "Some hours from here is a stilt village called

Laguneta. There you will find two parents with *el mal*." By now Tom and Nancy knew that the local term for Huntington's was *el mal de San Vito*, "St. Vitus's dance" or simply *el mal*, meaning "the sickness."

~ Looking for Homozygotes

To reach Laguneta, Nancy and Tom traveled three hours by jeep, then transferred to a boat. The boat, called a *chalana*, resembled a canoe with a motor. Lake Maracaibo was calm that day, but they had been warned it could quickly turn violent. "I think we're kind of stupid to be out here," Tom said, trying to sound cool and calm.

"I think you're kind of right," Nancy said. But they managed to reach the stilt village and come home safely.

The stilt village got its name from the tall wooden poles that support the houses, which were built right on the waters of the lake. In Spanish such places are called *pueblos de agua*, or "villages of water." The lakeside land was too marshy and junglelike to live on, so the villagers moved from house to house in boats. They supported themselves by fishing with nets on the rough lake waters. Children as young as six went out in the boats every day to fish. In the village were 22 small houses, a store, and a Catholic church. In each house lived as many as 18 people.

That particular day Nancy met a family that held a key to unlocking the secret of Huntington's, although she didn't yet know it. The father had been the chief of the village until the disease robbed him of some of his abilities. The mother, who was healthy, took great care of her nine children, seven of whom had the disease. (When HD appears in childhood, it has usually been inherited from the father.) Later, the DNA of a young son of this family would contribute crucial information to researchers.

But on that first trip Nancy was still searching for possible homozygotes for the defective HD gene. Through this boy's family, she found some. The boy's aunt (his father's sister) and her husband both had the disease. Even more important from Nancy's point of view, they had 14 children, several of whom showed symptoms of

Huntington's. But were any of them homozygotes? There was no way to tell. Any of the children with HD symptoms might have inherited the disease from just one parent or from both parents. None of those who were ill with HD were sicker than usual.

Still, Nancy and Tom explained to the parents and children that scientists in the United States and the rest of the world were trying to understand the disease better. The scientists could use blood samples from their family to gain critical understanding of Huntington's. Such knowledge could benefit not only them and their children, it could also help others suffering from HD in countries around the world.

The parents agreed to help. Soon after, Tom and Nancy boarded a plane to the United States with blood samples from the entire family. *This may lead to a small study but a useful one*, Nancy thought as she watched the ground drop away below their plane.

Back in the States, Nancy continued to think about the large number of Venezuelan villagers with Huntington's. The Lake Maracaibo region simply had to be an ideal place to try to unravel the mysteries of the disease.

"We need to go back to Venezuela," Nancy later said to scientists at a workshop on HD. "Where else can we find such large and

Heading down to a friend's house? In Laguneta in the 1980s, you had to pole your way there and back *(above)* because the houses in this stilt village *(top)* were built atop poles in the lake.

cooperative families with Huntington's who can teach us so much? They're the only people who can help."

Some scientists agreed with her; others doubted that accurate information on family relationships could be obtained in Venezuela or doubted that a young woman was the person to lead a team. But Nancy, stubborn as usual, was headed back.

In March 1981, Nancy and a scientific team were back in Venezuela. Several Venezuelan researchers joined the effort. Before they could find the HD gene, however, they needed to find a marker—a clue to where the gene might be.

~ Can That Be a Marker for Huntington's?

Genetic mapping is just like regular mapping, Nancy thought to herself. A map of the United States, a map of Los Angeles, or a map of New York are all divided by streets with names and even mountains and rivers. To find someone, you ask for their address. Finding a gene is all about real estate—who lives next to whom. Was our gene living on Madison Avenue? In Central Park? Or maybe in Santa Monica by the beach or in downtown Los Angeles? The problem was that there were almost no maps or signposts to the gene. The adventure was as daunting and exciting as Lewis and Clark setting out to map the unknown!

"Where else can we find such large and cooperative families with Huntington's who can teach us so much?"

So how do you find a gene? First, you try to find the gene's nearest neighbor. Nancy knew that big families with HD were needed because they will all have inherited their HD gene from the same person, called a founder.

This is how Nancy explains what they were looking for and why: "We needed lots of people in these families with HD symptoms, their parents, their grandparents, if possible, and also siblings and relatives who did not have HD. First of all, you need people with HD symptoms because you know they have the HD gene. You can see the symptoms in front of your eyes. Next, you need people in the same family—sharing the same DNA—who don't have HD symptoms. This is a little trickier because HD can appear as late as age 60 or 70, but mostly we can assume that family members with a genetic risk who were in their 50s and 60s and had not shown HD symptoms probably had inherited the normal version of the gene.

Homes on stilts in Laguneta sheltered parents and their children just as homes do everywhere.

"Again, we had to rely on what our eyes were telling us—who had symptoms and who did not—because we could not use a blood test to look for a gene we hadn't found. And we looked for families with both parents and four grandparents willing to participate in the study. Having many generations helps a lot, since you want to study who passes what gene to whom.

"Two genes sitting close together on a chromosome will tend to be inherited together. (Remember the penguins on the ice floe?) Knowing that the HD gene was passed down from generation to generation, we were looking for another gene that was sitting so close to the HD gene on the chromosome that it is passed down with it. Imagine there is a gene for making ice cream very close to the HD gene. This gene comes in 34 flavors. Suppose that, just by accident, a person with HD also has the chocolate flavor of the ice cream gene. That person's parent and grandparent and children and grandchildren with HD also have the chocolate flavor. Other people in the family, with the normal version of the HD gene, have vanilla, pistachio, strawberry—all different varieties but none have chocolate. This is very strong evidence that the ice cream gene is very close to the HD gene on the same chromosome. In other families, the flavor traveling with the defective version of the HD gene could be vanilla or mango. The ice cream gene does not have any effect on the HD gene. It just comes along for the ride.

"So how can you find out where the ice cream gene is, if you can't tell where the HD gene is, on which chromosome? Well, the ice cream actually melts. It leaves physical traces behind that allow you to sort out which chromosome it's on.

"The ice cream gene serves as a DNA marker for the HD gene. A DNA marker is a little variation in the DNA that comes in different varieties, or flavors. It has a precise home on a chromosome—just like a gene. If it is sitting close enough to the HD gene on the chromosome, it will travel with the HD gene from generation to generation."

Who needs a desk? Nancy pursues science wherever life takes her. A hammock slung on a Laguneta porch built on stilts is a fine spot.

~ Back in Venezuela

To introduce themselves and explain their work, Nancy and the team gave a big party for the villagers. The team included

geneticists, nurses, counselors, and doctors. They worked in three rural communities: San Luis, Barranquitas, and Laguneta. Nancy would return with a team every March and April for the next 22 years.

One of the team's first tasks was to construct a "pedigree" of the villagers. A pedigree is a family tree composed of everyone whose ancestry can be traced back to a common ancestor, along with husbands and wives who marry into the family and produce offspring. This work is critical because you need to know who gave their DNA to whom. It required lots of detective work to get it right. Nancy found out who was related to whom by

asking. There were no written records but, fortunately, the villagers often had a good picture of how they were related to one another and to deceased relatives. Even if the families knew the relationships, it was a challenge to translate this information into pedigrees and family structures. Lots of sleuthing and questioning were required.

Because so many people in the Lake Maracaibo area had similar names, names alone would not suffice for identification. The researchers took photographs of many people and pasted them onto the pedigree, which they then put up on walls. "Please stand near a photo of your closest relatives!" Nancy found herself shouting above the din of the crowd. With this unconventional method, she and her team sorted out who was who and constructed an enormous family tree of 18,000 people.

Nancy's sister, Alice, whose Ph.D. was in Latin American history, came to Venezuela with the team to investigate the history of the disease in this remote part of the country. Alice and team member Fidela Gomez proved that an individual thought to be the person who brought the disease to this area of Venezuela was, in fact, not the one. Eventually, thanks in part to Alice's research, Nancy's team

"A typical day in the office," says Nancy of this photo taken in Venezuela about 1990. Pedigree charts surround student Sheryl Lyss (left), Fidela Gomez (center), Judy Lorimer (standing), and Nancy.

succeeded in tracing Huntington's disease in the villagers back to one woman who had lived in a stilt village in the early 1800s. Appropriately, her name was Concepción. Perhaps her father was a European sailor with HD who had sailed to Lake Maracaibo, fathered children, and thus introduced the disease to this part of the world. No one really knows. What is known is that she had about 15,000 descendants in nine generations! In every generation since the time of Concepción herself, some of her descendants have had Huntington's. Many others have been (and are now) at risk.

Other families with HD do not trace their origins to Concepción but to other individuals with HD. The person who originates a family tree, who is the first in the family to have a disease and pass it on, is called a founder.

The team needed blood samples if the research was to go forward. At first it was not easy to persuade the villagers to give blood because most had never done so. The men feared they'd be unable to drink or fish if they gave blood. The children's eyes

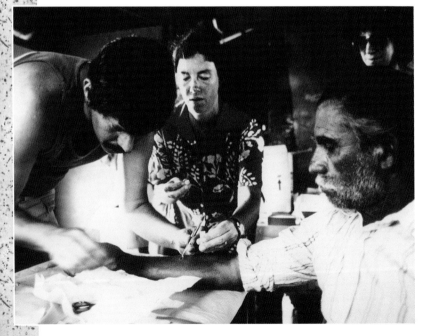

Ira Shoulson and Anne Young take skin samples from a Lake Maracaibo villager. Later, blood samples would be enough.

would grow wide with fright at the sight of all the tubes that had to be filled with their blood.

Nancy explained that she too came from a Huntington's family, meaning she was as much at risk as they were. She pointed to a scar on her arm that had come from giving a small skin sample and explained that previous researchers had used skin samples as well as blood for study, until they learned that blood would be enough. "See, I'm one of you," she said, moving through the crowd and showing the villagers her scar. "I've given my blood and skin to be studied to learn more about *el mal*. We are all one family."

64

Then she drove her point home: "Of all the people in the world, you are the ones who can help most to end this sickness. You are important to everyone everywhere who has it or might get it."

Until this time, the villagers had thought they were the only people anywhere to suffer the horrible fate of *el mal*. Thanks to Nancy, they now understood that they were members of a global family.

Once they understood, they gave their blood samples more willingly. Still, they found it hard to grasp that the disease existed in the United States as well. "If you can travel to the Moon," they said, "why do you put up with *el mal*? Why don't you just cure it?"

Eventually, thanks in part to Alice's research, Nancy's team succeeded in tracing Huntington's disease in the villagers back to one woman who had lived in a stilt village in the early 1800s.

For the Venezuelan blood to be useful for study, it had to reach Jim Gusella's lab in Boston within 72 hours after it was drawn. And because someone had to travel with the precious samples, blood was drawn only on "draw days," when one of the team members was about to leave Venezuela for Boston.

~ A Draw Day

It's another typical draw day in San Luis in 1982. The site is the blisteringly hot, poorly equipped government building that does triple duty as a school, clinic, and community center. The building and the village center are crowded because everyone in town wants to check out what these strangers are up to. Children cry; mothers comfort them. Men joke around and tease one another to mask their fear of giving blood. Tropical insects bite everything that moves—or stays still, for that matter. The walls are covered with the family pedigree chart.

Here comes Nancy in her khaki pants, cotton shirt, and sneakers, her long blond hair tied back in a ponytail to keep her neck cool. She hugs and kisses people, picks up the babies and small children, and speaks kind words to all in the basic Spanish

Nurse Fidela Gomez draws blood *(right)*. Anne Young and Nancy *(below)* enjoy a light-hearted moment together, as colleagues everywhere do, no matter how serious their work.

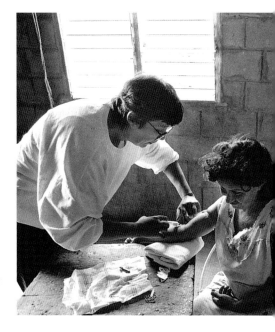

she's learned for these trips. She holds a tin cup filled with water to the mouth of a woman trembling because of HD.

Many members of Nancy's team are women. Everyone is very affectionate—family members and team members, both the men and the women on the team. "We touch them and they know we care. Also, they know now that we come back each year; we won't abandon them," Nancy says.

In a tiny examining room, Fidela Gomez, the nurse from Argentina, draws blood from the arm of each villager in turn. She is so quick, gentle, and amusing—and she collects each sample almost before anyone can be afraid. The team members take special care to label each blood sample accurately. By late afternoon all vials of blood have been collected and packed securely into Styrofoam boxes. Now one of the team members must journey by boat, jeep, and airplane back to Boston, where Jim Gusella is waiting at his lab at Massachusetts General Hospital to test the samples.

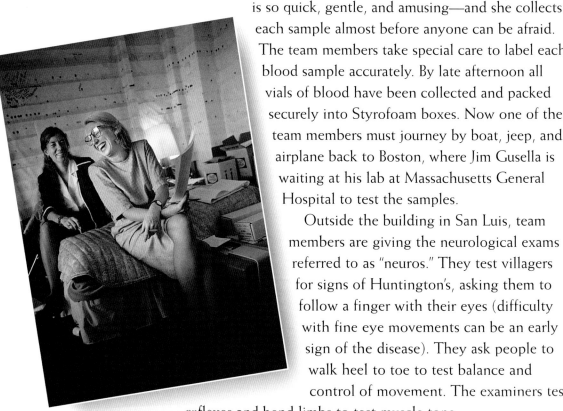

Outside the building in San Luis, team members are giving the neurological exams referred to as "neuros." They test villagers for signs of Huntington's, asking them to follow a finger with their eyes (difficulty with fine eye movements can be an early sign of the disease). They ask people to walk heel to toe to test balance and control of movement. The examiners test reflexes and bend limbs to test muscle tone.

66

The exam reveals who is ill and, in those already diagnosed, how far the disease has progressed since last year. The children make a game of the neuros, pretending to be doctors and patients. They do excellent imitations of both groups.

Though the researchers use the neuros to test for symptoms of HD, the villagers themselves are very good at diagnosing the disease. When they say that someone is *perdido* (lost) when he or she shows even slight symptoms, they tend to be right.

Many of the team members have come every year since 1981, enduring great hardship in the hope of someday helping others. Anne B. Young, chair of neurology at Harvard Medical School, her husband Jack Penney, and Robert Snodgrass of Children's Hospital in Los Angeles are some of the extraordinary team members who have continued to make regular visits.

Pediatric neurologist Robert Snodgrass greets an elderly gentleman in Laguneta. A dedicated member of the team since 1982, Bob has always shown great admiration and respect for the family members.

Although there is no treatment for HD, the team doctors and nurses do treat the villagers for other ailments. Some have injuries; others have infections or even parasites.

On another day the team travels to Barranquitas, two hours south of San Luis. In this village, too, the way of life has not changed much for hundreds of years. Here, at least half the population is sick with Huntington's or is at risk of developing it. More than 3,000 children in the region will eventually become ill and die of the disease. But today, groups of curious children surround the team members as they unpack their equipment. Nancy introduces a new doctor on the team to one of the children who seems healthy, a little girl with big blue eyes and a quick smile. Little Teresa had become almost an adopted daughter of Nancy's team.

To reach Laguneta, the stilt village south of Barranquitas where most inhabitants are at some degree of risk for HD, the team travels first on an oil company ferry, and then on small fishing boats. The researchers make their way from house to house by boat, intent on visiting every family. At night they sleep in hammocks slung across porches outdoors. Bats and jungle insects are never far away.

In all the villages, Nancy and her team do neuros, draw blood, treat medical problems, and keep updating the pedigrees. "We never stop asking questions about family, about how each person is related to the others," she explains to the new doctor. "We come back year after year, reexamining people to record how the disease has progressed." Over time, many team members grow so emotionally

Colorful plates and bowls may liven up a village house on the water, but it's the family members who make it a home.

close to the villagers that they must fight depression and despair as each return trip reveals the inexorable march of the illness.

In every village, Nancy's nickname is the same. Every March people look forward to the return of "the blond angel."

One year it seemed the blond angel and her crew might never return. That was the year a terrible storm blew over Lake Maracaibo, catching Nancy and the others on a boat in midlake, traveling from one village to the next. For hours they battled the wind and the storm-tossed water, keenly aware that many people had drowned on the lake. When they finally arrived safely, they held a big dance and everyone in town came to celebrate—not only the villagers who were well but the afflicted ones who couldn't stop their strange dancing.

Nancy Wexler took *pride*
in the fact
that she had helped devise

a way for people at risk
to divine their *futures*.

TESTING THE FUTURE

7

In the early 1980s Nancy was in her late 30s. Because Huntington's strikes most often when its victims are in their 30s or 40s, she worried whenever she dropped a pen, tripped, or forgot an appointment. Was this the onset? Would the same things now happen more often?

Then she made herself stop worrying and get on with her work. She'd found by experience that worrying didn't help her when she was powerless to change things. Work, on the other hand, could change the future—hers and others'—and was much more fun than worrying!

Nancy's work kept her in Washington. When the Congressional Commission for the Control of Huntington's Disease ended, she joined the National Institute for Neurological Diseases and Stroke (NINDS) at the National Institutes of Health, in part to carry out the commission's recommendations. She also remained deeply involved with the research on blood samples in Boston, which was slowly but surely moving forward.

David Housman was an optimistic scientist who encouraged Nancy and the Hereditary Disease Foundation to push forward using DNA markers to find the HD gene. David advised testing each new marker developed to see if it was close to the HD gene, instead of waiting until the entire human genome was mapped.

Most science progresses with a mix of meetings, writing and reading papers, and hands-on lab experiments *(opposite)*. The pedigree or family tree *(above)* contributes vital data for HD research.

David also strongly encouraged Nancy to go to Venezuela to do the research. Nancy naively asked him if bigger is better and David answered, "Definitely! The biggest family you can possibly find is best!"

David had a terrific graduate student named Jim Gusella. Gusella and his team moved from M.I.T. to Harvard University and were busy developing new markers that could be used to help pinpoint the location of genes. These markers were called restriction fragment length polymorphisms, or RFLPs (sometimes pronounced "riflips").

RFLPs are small, normal variations in DNA that serve as markers or signposts. Each RFLP has a specific home or address on a chromosome, just like a gene. Each new marker Jim and his group developed was tested to see if it "traveled" with HD in a family. From late 1980 to 1983, Jim fashioned 11 new markers, but none hit pay dirt. Were the pessimists right?

Jim Gusella's team included a technician named Ginger Weeks. She designed the 12th marker and they named it after her: G-8. G-8 came in four different varieties, or flavors: A, B, C, and D. If the G-8 marker was sitting close to the home of the HD gene on a chromosome, G-8 would be inherited

Dr. Michael Conneally is a well-known population geneticist at Indiana University. His team did the computer analysis that helped search for the marker for Huntington's disease.

consistently with the HD gene. Jim and his team analyzed the DNA of the families with HD and then sent their data to Mike Conneally at Indiana University and his team of statistical geneticists. Mike combined this DNA information with Nancy's pedigree and symptom information.

Jim, Mike, and Nancy first analyzed the data from a large Iowa family with HD. The analysis hinted at the possibility that G-8 might be close to the HD gene. But when Mike put everything into the computer, the results were not statistically significant. The data didn't show a clear connection.

Jim and his group were very busy extracting DNA from the blood cells Nancy and her team had been collecting and sending up from Venezuela. Jim took this DNA data together with Nancy's Venezuelan pedigree information and results of the neurological exams and sent it off to Mike Conneally to be analyzed. And then he went to a meeting in Aspen, Colorado. Nancy stayed in Washington, D.C., pacing back and forth frantically. What would happen?

They were all waiting to see if G-8 was close to the HD gene. If it was close, the HD gene would be sitting near one of the four variants. By chance it could be A, B, C, or D. If G-8 and the HD gene were close, then the defective HD gene would travel with one of the variants and the normal version of the HD gene would travel with another variant.

Everyone had practically stopped breathing, waiting for the results from the Venezuelan samples. Were they on the eve of a monumental, world-shaking discovery or a huge disappointment?

Mike was about to leave for a vacation in the Grand Canyon when the G-8 data arrived. "You guys feed this data into the computer," he told his graduate students, Peggy and Beth. "Then let me know what you find." And he took off with his wife, son, and daughter, heading west.

Mike was staring into a campfire at Grand Canyon National Park when a ranger came looking for him. "If your name's Conneally," the park ranger told him, "you're supposed to call Peggy and Beth."

It was late—11 P.M.—so Mike figured the morning would be plenty soon to call. How could he know the news was so urgent that Peggy and Beth would stay up all night waiting to hear from him?

Nancy was apparently right when she noticed that doing research made scientists happy. At the National Institutes of Health, she relished her work researching HD inheritance patterns.

~ *Eureka!*

In the morning, Mike learned from a groggy Peggy and Beth what the computer analysis had revealed: Astonishingly, the 12th marker made by Jim's group and the only one tried in the Venezuelan families had hit the jackpot! Amazingly, everyone who had HD symptoms in the Venezuelan families had the C form of G-8 and their normal relatives had A, B, or D. The odds of this happening by chance were over a million to one.

Nancy, waiting to hear, was desperate. She knew Mike was at the bottom of the Grand Canyon and unreachable. She couldn't reach Jim in Colorado. Finally, no longer able to bear the suspense, she called Peggy and Beth and learned the fantastic news! She screamed so loudly at the good news that everyone came running to find out what was wrong. Nancy first called her father and then her sister. Milton was ecstatic. He said, "I am glad I lived to celebrate this day!" To Alice, Nancy said, "It was total joy. I was on the ceiling, jumping up and down!"

Michael Conneally, Nancy, Milton Wexler, and Jim Gusella pose with a celebratory cake after the HD marker is found in 1983.

Then Nancy ran up to her colleagues at NINDS to thank them for making this all possible—funding the Venezuela research and paying for all the things she needed for her work, from medical equipment to rented motorboats.

As for Mike Conneally, he exclaimed, "When I heard the news that the search was over I was almost crying."

Next, scientists narrowed in on the general location of the gene. They could place it somewhere on the top of the short arm of chromosome 4 (of the 23 chromosomes). But finding the gene's exact location would require years more of hard work.

News of the marker and its location on chromosome 4 hit the world in November 1983, when Nancy, Jim Gusella, Mike Conneally, and others published a paper about it in the British science journal *Nature*. This was the same journal in which pioneering geneticists James Watson and Francis Crick had unveiled the structure of DNA.

Almost every major U.S. newspaper ran the story on its front page. *Time* and *Newsweek* covered the story, too. Nancy and Jim appeared on TV and radio shows, telling the story of HD, its victims, the extended Venezuelan family, and their breakthrough in localizing the gene. Jim Gusella could only shake his head in wonder as he remembered, "When I first decided to work on this research, people thought I was just plain nuts."

What would the new finding mean to people like Alice and Nancy who were at risk for the disease and to those who were already ill? There was still no cure, nor even any truly useful treatment. Doctors could offer some relief for certain symptoms, such as depression, but no one knew how to stop patients from sliding down the slow and painful path to death. Still, Nancy and her team knew that the marker discovery now, the chance of locating the gene itself soon, and the hope for a cure someday was the best timeline they could offer HD families.

> Who would willingly chance hearing the chilling news, "You have the gene for Huntington's disease—and because there is no cure, you will get sick and die of it."

Clearly, a revolution had occurred. Now that there was a marker, people at risk for Huntington's had a devastating decision to make: They could choose to take a new test that would tell them, with 95 percent accuracy, whether they had inherited the HD gene. (Today's test is closer to 100 percent accurate.)

Who will take such a test? That was one question. This test must be totally voluntary, given the terrible implications. Another, equally vital, question was this: Who will want to take it? Who would willingly chance hearing the chilling news, "You have the gene for Huntington's disease—and because there is no cure, you will get sick and die of it."

~ Do You Want to Know?

When the chance to be tested was first offered, it was not enough to test the blood of only the person at risk; the test involved the whole family. There had to be many living family members who were available and willing to be tested. Nancy explained the situation this way: "Within any given HD family, the same pattern of the marker tends to travel with the gene. The marker patterns of all the relatives must therefore be traced to find out which of the four versions of the marker travels with the appearance of the HD gene in that family."

People who had been adopted or did not have many living relatives could not obtain useful information from the test. But those who did have enough willing relatives had to decide whether they wanted to know if they carried the abnormal gene.

Carefully, she laid out the benefits and drawbacks of taking the test.

As a psychologist, Nancy was vividly aware of the emotional dangers the test presented. Yes, it would be wonderful to hear the news that you would never get Huntington's. But did that chance offset the odds of receiving what amounted to a death sentence—the news that you were certain to get the disease?

Nancy counseled those at risk, as well as their spouses and other family members. Gently, she explained both the science and the psychology involved. Carefully, she laid out the benefits and drawbacks of taking the test. Children should not be tested, Nancy counseled, because they were too young for such a monumental decision. Nor was it right for others to invade their privacy by having them tested. Once a test had been given, Nancy or other counselors always gave the results in person.

The test could be useful for young adults planning to have children. If the test revealed that they did not have the gene for HD, they could rest easy, knowing their children would never get the disease either. If the test revealed the opposite—that they did carry the gene—they faced a complex decision: Should they have children of their own, knowing the chances were 50 percent that they too would develop the disease? New in vitro fertilization techniques now make it possible to implant embryos that do not carry the abnormal HD gene. But the parent must still decide whether to have children, knowing the disease will eventually strike the parent, possibly when his or her children are still young.

~ Decision Time

What would the Wexler sisters decide about testing? They met for lunch in Los Angeles to talk over the problem. Nancy said, "I always thought, sure, I'd want to be tested. I never had any doubt."

"But what do you think now?" Alice asked.

"Now I think it's like watching someone outside myself," she replied, "wondering what she will do." Neither sister reached a decision that day.

The next weekend the sisters met with their father to discuss the test. Of course, Milton had always hoped his daughters would not get the disease. Now the reality of what they might learn sank in on all of them like lead. "If either one of you has the abnormal

Nancy and Milton have collaborated on HD throughout her career. Often it's workshops like this one organized by the Hereditary Disease Foundation that bring them together professionally.

gene," he said, "all three of our lives will be ruined. Why risk destroying everyone's happiness with this knowledge now?" The issue was so emotional that logical discussion seemed difficult.

For the time being, the Wexlers dropped the subject. Whatever decision the sisters ultimately made would remain a private matter.

Maybe it was a case of the old proverb: "Beware what you wish for; it might come true." Nancy had wished for—and worked hard to develop—a test. Now she and those closest to her had to deal with the reality that such a test existed.

Whether she herself eventually decided to take the test or not, Nancy Wexler took pride in the fact that she had helped devise a

With pedigree data and blood samples, Nancy pursues her passion to save the lives of many thousands of people worldwide destined—by a mistake in their DNA—to die of Huntington's disease.

way for people at risk to divine their futures. But the real benefit of having the gene in hand is to lead to better understanding of the disease, bringing prevention, treatment, and a cure.

~ Testing, 1, 2, 3

In the second half of the 1980s, testing centers for Huntington's opened in the United States, Canada, Great Britain, and several other European countries. At every center those who request testing are asked to participate in a program designed by HD families and those who work with them. The program includes intensive counseling as well as a neurological exam to see if they are showing any HD symptoms. Counseling sessions focus on questions such as these: Do you really want to know, and why? How do you think it might change your life to know? Can you accept a bad result and go on living your life? Will you have "survivor guilt" if you have the normal form of the gene and others are suffering? Will you return for more counseling if you need it?

Once a person made the decision to go ahead, technicians drew blood. The blood samples were then analyzed for the DNA markers close to the HD gene. But in every country, far fewer people at risk requested the test than had been expected. By 1992, only 1,400 people at risk in the entire world had taken the test. In fact, less than 20 percent of adults at risk come to one of the testing centers. Like the Wexlers, people everywhere found the testing decision to be the hardest one they had ever faced.

~ "Live the Way You Want to!"

The technique of testing for HD has become easier since it was first introduced in the 1980s. All that's needed now is a DNA sample from the person who wants to know his or her risk status.

The test itself may be simpler, but deciding to take it remains as difficult as ever because there is still no treatment, prevention, or cure for HD. Nancy and other psychologists have counseled many at-risk people struggling with the decision. People may want to use the test to end uncertainty about the future, but if they find they have the abnormal form of the gene, they are opening a new Pandora's box of uncertainty. When will the disease appear? How old will I be? Will it be severe? Mild? Will I turn into my mother or father?

When Nancy lectures or presents papers about Huntington's disease at conferences, she emphasizes that people do not realize how difficult it can be, when you are young and healthy, to face the fact that you will die of a catastrophic illness and there's no way to stop it. Even if you ponder and think very carefully about whether or not you want to know your future, you can't really imagine how you will feel after gazing at the crystal ball and having a true fortune told. Experience has taught most psychologists that people who take the test must be prepared to get depressed if the results come back positive. Nor are those who find out they don't have the abnormal form of the gene as happy at that news as they thought they might be. They still worry about whether their siblings carry it. They may have to cope with feeling guilty that they are healthy when those they love may not be.

> Regardless of age, gender, or any other factor, choosing to be tested is a heavy decision for anyone at risk.

Other factors, too, influence the decision. People at risk may rightly fear losing their health insurance someday if they test positive. Or they worry with reason that their employer won't promote them. Or they're afraid, with cause, that no one will want to date them or marry them.

Many people have confided to Nancy that they wanted to be tested so they could decide on some major step in their lives, such as preparing for a new career. Or they've said, "If I knew I was going to get sick, I'd take my family on a cruise now." Nancy's advice—and that of most psychologists—is usually, "Why not live the way you want to live, whether you have the Huntington's disease gene or not? Go to school, change jobs, take that family trip. Live your life!"

For younger people at risk, a weightier decision is whether to have children. Although some people at risk choose to adopt children rather than take a chance of passing the disease to the next generation, one option, for couples who feel it's the right thing to do, is to test the fetus during pregnancy. Another available option is preimplantation genetic diagnosis which involves testing for the presence of the HD gene in embryos that have been created in the lab by combining the parents' eggs and sperm. Only embryos that are free of the Huntington's disease gene are then implanted in the mother's womb.

More women choose testing than men. More older people than young ones take the test. Very often, older people get tested if they have children. They might prefer not to be tested for their own sake, but they get tested to clarify the genetic risk for their children and grandchildren. If the parents are free, their descendants are free. If they are found to be carrying an abnormal gene, their children have a one-in-two risk of following in their footsteps.

Regardless of age, gender, or any other factor, choosing to be tested is a heavy decision for anyone at risk.

The **hunters** may have captured the gene,
but the cure remained
the next elusive **prey**.

Without a pause, the hunt continued.

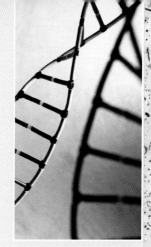

8

WE FOUND IT!

I n 1984 Nancy moved back to New York City, where she had
been appointed a professor at Columbia University. She quickly
realized she needed someone to work with: to direct and manage
the Venezuela project and also to help organize her research and
teaching and her monthly visits to Los Angeles to see her father and
sister and to work for the Hereditary Disease Foundation. On top of
all that, each year she traveled to conferences all over the world.

Nancy had recently run into Michael "Tiger" Lorimer, her
guitar-playing friend from high school. Michael had introduced
Nancy to his wife, Judy, who was looking for a job after managing
Michael's concert career around the world.

The demanding boss obviously needed a collaborator unafraid
of a difficult job. Fortunately, Judy Lorimer wasn't put off by the
challenge. She took the job, and before long Nancy was telling
other people that "Judy is phenomenal."

Michael, too, became involved in the fight against Huntington's
disease. He gave several guitar concerts to raise money for research,
and he sold CDs of his music to benefit the Huntington's community.
In a nod to his old high school nickname, the CDs were produced
under the "Tiger Tunes" label.

Nancy (opposite)
wants more than
anything to find a
way to stop chil-
dren's futures from
being stolen by HD.
The DNA helix
(above) features a
"twisted ladder"
form.

~ Operation Cooperation

While the world celebrated the birth of a brand new strategy for finding genes, the Wexler family and the scientists realized that with only a marker and no HD gene, they were nowhere. Hard as it was to find the marker, it was even more difficult to find the gene!

The Hereditary Disease Foundation convened a workshop in 1983 to confront this hurdle. Jim Gusella, proud discoverer of the G-8 marker that narrowed down the location of the HD gene, was there as was David Housman, Jim's former mentor; Jim and David had originally had the idea to use DNA markers to find the gene. John Wasmuth of the University of California at Irvine was also there. John had developed a special human chromosome that had the tip of chromosome 4 attached to the rest of chromosome 5. John realized that the HD gene was in this piece of chromosome 4 that he had captured on the top of chromosome 5 and this further

The "Gene Hunters," including Nancy *(far left),* and advisory committee members pose together during a working vacation at Dennis Shea's retreat in the Florida Keys.

narrowed the territory needed to study. The trick was to limit the region you had to study on the top of chromosome 4 to as small an area as possible, without eliminating the HD gene.

Hans Lehrach, from the Imperial Cancer Research Center in London, had a clever way of trying to slice off the top of chromosome 4 mechanically. Francis Collins was a newly minted geneticist with bright ideas. Peter Harper (now Sir Peter) was an HD specialist and medical geneticist from Cardiff University in Wales.

These were the principal investigators (or primary scientists) of six international laboratories who worked zealously for a decade to find the HD gene. Over this time, many graduate students, postdoctoral students, and technicians in these labs labored round-the-clock with no weekends off to reach the goal as quickly as possible, since lives were at stake. Officially called the Huntington's Disease Collaborative Research Group, the group was nicknamed the "Gene Hunters."

Funding from the Hereditary Disease Foundation supported work in the labs. Members of the HDF's Scientific Advisory Board generously met with the group to help guide the scientists' research.

These advisors were: Bob Horvitz, of M.I.T., who went on to win a Nobel Prize in 2002 for his discoveries about how cells die; Richard Mulligan, of the

> Over this time, many graduate students, postdoctoral students, and technicians in these labs labored round-the-clock with no weekends off to reach the goal as quickly as possible, since lives were at stake.

Whitehead Institute at M.I.T., who had won a MacArthur Genius Award in 1981; and P. Michael Conneally, who had participated in the team finding the HD marker.

At that time, people were just inventing machinery to look at DNA. It was only a short while after sequencing—learning how to look at each base pair in the double helix—had been invented. The machines the Gene Hunters used were mostly rigged together with large wads of paper towels and held together with giant clips!

Certain postdoctoral students shone in this group and have continued to work on HD to this day, even after the gene was

British scientist Gillian Bates *(right)* inserted part of the human HD gene into a mouse's DNA, creating the first HD mice.

found. In particular, Gillian Bates, who worked with Hans Lehrach, made the very first mouse with the human HD gene in it. Leslie Thompson, working with John Wasmuth, made the first fruit fly model of HD. And Marcy MacDonald, working with Jim Gusella, continues to figure out how the HD protein does its damage.

The scientists decided to publish as a group so that no one would be left without proper credit after putting in a decade or more of hard work. Also, the work was exceedingly, excruciatingly difficult and they needed each other.

The group actually invented 14 new technologies for handling and manipulating DNA and for finding genes. Along the way, they helped other researchers or they themselves found genes for cystic fibrosis, breast cancer, Elephant Man's disease, Lou Gehrig's disease, Alzheimer's disease, and many others.

As a group, this forward-thinking, imaginative, and creative team helped inspire, shape, and formulate the Human Genome Project, aimed at finding and understanding all our genes. When in 1983 HDF scientists Nancy Wexler, Jim Gusella, Mike Conneally, and the team discovered the marker that was genetically linked to the HD gene, part of the "Eureka!" that was heard around the world was the realization that these same techniques could work for finding any gene. Discovering the Huntington's disease gene was the launching pad of the Human Genome Project.

Planning for the Human Genome Project and even beginning the project was simultaneous with looking for the HD gene. Many of the Gene Hunters contributed their skills to the service of discovering all our genes. In fact, Francis Collins went to Washington to head the National Institute for Human Genome Research at the National Institutes of Health. This new institute was created to direct the public human genome effort. Nancy also played a crucial role in this effort, chairing the Ethical, Legal, and Social Issues Committee of the Human Genome Project.

It's rare for top scientists to agree to work together, sharing not only their results but also the credit for any success. Everyone agreed that Nancy was the catalyst responsible for putting together this unique collaboration.

Everyone in the group recognized they owed a debt to the people of Venezuela. Without being able to study the DNA in blood donated by the villagers, none of the research results so far would have been possible. So as the search for the gene continued, so too did Nancy's visits to Venezuela. Every March she packed her supplies and made the long journey south to the villages dotting Lake Maracaibo. Judy Lorimer, the trip's director, always came along, as did Julie Porter, in charge of data. Herb Pardes, Nancy's partner, often did, too. On each visit the team took pedigree information and performed cognitive assessments and "neuros" on those they had examined the year before. Had symptoms appeared that were absent last year? Was the disease more advanced? The villagers themselves made new diagnoses, which they passed along to the research team.

An older Michael "Tiger" Lorimer with his wife, Judy, Nancy's invaluable, "phenomenal" collaborator and friend.

One year when the team arrived, the scientists looked eagerly for 15-year-old Teresa. There she was with her mother, laughing, happy to see Nancy and the others and looking the same as ever. But her mother was crying. "My Teresa is lost," the mother said. There again was that phrase of finality the villagers always used: *perdida*, or lost to *el mal*.

Nancy hugged Teresa. Afterward, she remarked to Bob Snodgrass, the neurologist, "I could feel her twitching."

"We'll see," he said. "She doesn't seem sick to me." But after he had carried out the customary neurological tests on Teresa—asking her to track a moving finger, checking her reflexes—he went off by himself and cried.

The villagers never seemed to err in their diagnoses.

~ A Story Made for TV

In March 1986 a crew from the television show *60 Minutes* came to Venezuela to film Nancy and her team at work with the villagers. The program's producer recognized the huge audience appeal in the story of a scientist and her family at risk for a mysterious disease and all fighting for their lives.

Audiences would be especially interested in the question of testing—and not just for Huntington's. When testing became available for other diseases, millions of people would face critical decisions. TV viewers would want to know how Alice and Nancy felt about testing: What's it like for people at risk to make these choices?

In April, when Nancy and Alice returned to Los Angeles, host Diane Sawyer and her team of assistants and camera operators filmed the rest of the *60 Minutes* segment, including scenes on the beach and some in Milton's apartment. Sawyer discussed the testing issue on camera with all three Wexlers. Nancy repeated what she had told her sister: "I was always positive I'd want to be tested. But now it's the hardest choice I've ever had to make in my life." Neither sister revealed what she planned to do.

> The program's producer recognized the huge audience appeal in the story of a scientist and her family at risk for a mysterious disease and all fighting for their lives.

Diane Sawyer asked, "What is your fantasy about what you would do on the day after a cure was found?"

Nancy answered, "I'd go door to door in those villages in Venezuela—where thousands of kids are going to die of the disease—and I would pass out pills to them. Everyone would cry for joy. I can't imagine anything more wonderful!"

~ Margot, Mangoes, and Miracles

But that joyful day might be far in the future, and the work in
Venezuela went on. In 1991 a Venezuelan doctor named Margot
Mieja de Young began working with Nancy's team in the villages.
One day she observed a family trying to deal with a relative whose
Huntington's caused a disturbed mental state. He had climbed a
tree and refused to come down. "I'm a mango in a mango tree!" the
man declared.

Enter Dr. de Young in her bright print shirt and turquoise
baseball cap. "That's wonderful!" she called, smiling up at him.
"But you are ripe now and it's time you were on the ground."
Without a word the man climbed down the tree and went on
his way. Problem solved.

Nancy could not praise Dr. de Young enough. "Margot is a
miracle worker. She is perfectly attuned to the people emotionally."

Dr. de Young had no cures to offer for HD because there were
none. She treated everyone in town for other ailments like broken
bones (HD patients fall a lot), pneumonia (a common illness in
Huntington's), and lice and parasites (a routine hazard in this part
of the world).

Dr. Margot Mieja de
Young feeds a young
girl stricken with
Huntington's who lives
in the Casa Hogar. The
team always provides
loving care along with
medical expertise.

Dr. de Young and the other doctors and researchers who joined Nancy every year in the Lake Maracaibo region performed heroically. But Nancy was aware that a critical element was still lacking. They needed a permanent place to care for the villagers. In 1992, thanks to constant pressure from Margot and Nancy's team, the Venezuelan government and a group called Friends of People with Huntington's Disease bought El Toro Rojo, or "The Red Bull," the roughest, rowdiest bar in the Lake Maracaibo village of San Luis. The plan was to turn the bar into a clinic and research center. Where once there were only drunks, drug addicts, and brawls, there would be a medical staff and caregivers devoted to HD-stricken villagers. But it was over a decade before enough money was available to rebuild and equip the old bar.

It took a decade of hard work and fund-raising by Nancy and others before patients with HD had a home at the Casa Hogar.

~ Victory!

When not in Venezuela for her annual working visit, Nancy immersed herself in the search for the HD gene. Not only was she one of the researchers, she was also the cheerleader for many other scientists. By 1989 some 50 researchers from 10 different institutions were members of the Huntington's Disease Collaborative Research Group, on the search for the HD gene.

Why did so many scientists dedicate themselves to working on HD? After all, the illness affected only a fraction of the world's population. One reason was their confidence that solving the mysteries of HD would shed light on the workings of many other diseases, particularly genetic (or inherited) diseases and neurodegenerative diseases (those in which brain cells die). This knowledge in turn might allow those diseases to be conquered.

Many of the scientists had visited the villages around Lake Maracaibo or had watched videos of people from the area. They had seen children with the rigid gait that signifies juvenile HD— little children, old and dying before they had blossomed. They had seen adults stumbling their tortured way down village paths. For these reasons the search for the gene was intensive.

> That lifelong determination, they now knew, had led Nancy to crack the HD nut.

By 1992 the group had homed in on "candidate genes"—about 100 genes residing in the region of chromosome 4 where the scientists expected to find the Huntington's gene. One of these genes might well be the "crown jewel" the group was seeking— but which one? An enormous amount of work would be required to single it out.

Then, in February 1993, the long-awaited news broke: The Gene Hunters had spotted their prey—the gene that causes Huntington's disease. Ten years of unique scientific cooperation had finally paid off!

Even the researchers who had done the work were amazed. One claimed he was "literally stunned." Relief, happiness, and disbelief—these emotions circled the globe along with the good news. Newspapers everywhere carried the story.

"I'm ecstatic to see this day. Bravo to the dedicated people who stuck with it for a decade!" said Milton Wexler, who turned 85 that year. His daughter's insistence that the gene could be discovered— that it would be discovered—had fueled the search. Nancy's father and sister remembered the little girl who had stubbornly insisted on dressing herself. That lifelong determination, they now knew, had led Nancy to crack the HD nut.

~ Connecting the Gene to the Disease

How does the Huntington's gene cause the disease? The gene makes a protein called "huntingtin." This protein contains an amino acid called glutamine. In HD patients, extra glutamines result from too many repeats of the following sequence of chemical building blocks in the DNA: C-A-G (the letters stand for cytosine, adenine, and guanine). Neurons in the brain are somehow destroyed by those extra glutamines. Because the neurons have been destroyed, the brain no longer works properly and symptoms of Huntington's disease begin to be observed.

Angel, a child from a Venezuelan stilt village, lived only 11 years, but his contribution to HD research will benefit people around the world for generations to come.

~ A Little Boy's Legacy

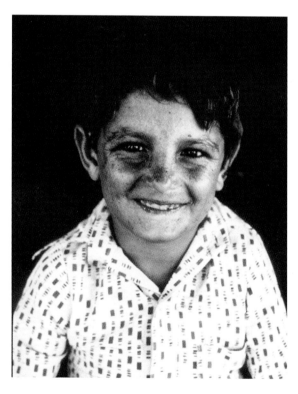

Much of what the scientists learned about repeating CAGs was the legacy of one little boy in a faraway village in Venezuela. When Nancy and Tom Chase first visited the village, they met the little boy, whose father had been chief until he was disabled by HD. His son, Angel, also developed the disease at age 2 and died of it at age 11, but his DNA was a huge help in identifying the gene. No one associated with HD research will ever forget the contribution of this child, his immediate family, and the thousands of members of the pedigree who were his extended family.

What was it about the little boy's DNA that led to identifying the HD gene? Jim Gusella had been looking at DNA from the child's family in his lab when he noticed something on the boy's DNA that looked like a spot or a clump. At first Jim thought he was seeing a marker. Further analysis, however, revealed that he was looking at an expansion of

Too Many Repeated CAGs

Every gene is made up of a string of chemical building blocks, or nucleotides, in some order. It's the order that defines the gene. These nucleotides—thymine, cytosine, adenine, and guanine—are represented by an "alphabet" of four letters: T, C, A, and G. When DNA contains the sequence CAG (cytosine, adenine, guanine), inevitably these are instructions for the cell to make the amino acid glutamine. The normal gene for producing the huntingtin protein has up to 35 CAGs or glutamines in a row. When the sequence repeats more times than that—an event dubbed an expansion of the DNA—Huntington's disease results. People with more than 40 CAGs in a row get the disease. Moreover, the greater the number of repeated CAGs, the earlier the person inheriting the HD gene will begin to show symptoms.

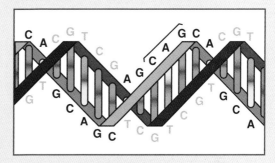

This small section of DNA shows six CAG repeats. Forty or more repeats result in HD.

the DNA. Indeed, Jim found over 100 CAG repeats—that's a very large expansion—in the boy's DNA. That large expansion was the clue researchers needed to locate the HD gene.

~ Celebrating a Milestone

About a month after the HD gene was found, Nancy and her team returned to Venezuela. They threw a big party to celebrate the scientists' breakthrough discovery. The party took place at the Red Bull, the former bar that was being transformed into a clinic. Everyone enjoyed the music, food, soda, and candy. But when the villagers understood that there was still no cure, they asked doubtfully, "Why are we having a party then?"

"Science moves in steps," Nancy explained. "Every step is important. This is a big step and it is worth celebrating."

The hunters may have captured the gene, but the cure remained the next elusive prey. Without a pause, the hunt continued.

The Venezuelan families
have joined hands across the world

with Nancy's family and others to

find the cure

for Huntington's disease.

IN QUEST OF THE CURE

9

Nancy picked up the small mouse that was busily grooming itself and stroked it gently. She was at Guys Hospital in London, visiting Dr. Gillian Bates at her laboratory. Gill had been a critical member of the Gene Hunters team when she was a postdoctoral student with Hans Lehrach. Now that they had discovered the HD gene, she could put that gene into a mouse and create the first mouse with the actual human gene. This meant that she and others could develop and try new drugs on such mice before testing the drugs on patients, to see if any would work. Gill and others could also learn from these mice more about the human disease itself.

The mouse in Nancy's hand today had HD, or as close as a mouse can come to having the human disorder. Gill had set out to create a mouse model of the disease, but the task had been enormously complex. Animals don't naturally have HD, and before the puzzle of HD can be put together, the disease must be understood in animals that are as close to humans as possible. You can't try a new drug in people without checking it first in mice. How could she make a mouse that exhibited the uncontrollable movements of Huntington's? Which part of the enormous HD gene would cause the mouse to twitch and seem to dance? Would the gene affect the animal's mental abilities, too?

Nancy, in a typical Venezuelan setting, is intent on gaining information to advance the science of HD *(left)*. In the laboratory, mice carrying the HD gene may help scientists find a cure *(above)*.

Gill's first step was to create HD mice. She did this by inserting part of the human HD gene into the mouse's DNA. She also placed 150 CAG repeats into that gene fragment. The result was a group of transgenic mice; that is, each mouse carried a gene from another source.

And what happened? Between the ages of 3 and 21 weeks, the transgenic mice developed symptoms similar to those of human HD patients. They shook and trembled. Their wobbly walk resembled the dancelike motions of Huntington's. They ate a lot but stayed thin. They made the same movements over and over, such as petting their noses or grooming their fur. In fact, when Nancy returned the mouse she'd been stroking to its cage, it went right back to grooming itself. These mice even had learning and memory problems.

The transgenic mice became increasingly ill. Gill observed that the mice with the greatest number of CAG repeats were severely ill after three months (a normal mouse lives two years). The ones with a smaller number of repeats might last many more months. In addition, their symptoms were less severe. Continuing to work with her mice "offspring," Gill and her colleagues succeeded in identifying huge clumps of the HD protein in the nerve cells of transgenic mice and humans with HD.

~ Learning from Mice

Huntington's researchers found mice to be valuable assistants throughout the 1990s. Two Columbia University scientists, Argiris Efstradiadis and his postdoctorate student Scott Zeitlin, created what they called a "Knock-Out" mouse. They removed the mouse's own version of the HD gene but did not replace it with anything else. (The mouse version is 87 percent identical to the human HD gene.) The result? The mouse died, demonstrating that the HD gene performs a function that is necessary for life. It's probably impossible to cure HD in people by removing their HD gene. The HD gene seems to make a protein that's critical for life.

The same scientists created a "Knock-In" mouse by enlarging the size of the animal's own HD gene. These mice displayed subtle symptoms of Huntington's. Later, another "Knock-In" mouse was given a huge CAG expansion. The result was severe HD symptoms that started early in the animal's life.

Then there's the Mouse House, established by Flint Beal at Weil Cornell Medical School in New York. Like Gillian in London, these scientists test drugs that might prove useful in treating Huntington's, in several different mice models of HD. At the same time, in laboratories around the world, scientists were experimenting with various drugs on HD mice to see how the drugs affected their behavior, symptoms, and life span.

One experiment was especially exciting in mouse research. Ai Yamamoto, a graduate student with Rene Hen at Columbia University, put the human HD gene in a mouse with a special genetic "switch," a way to turn off the gene after the mouse started showing symptoms. Amazingly, the mouse's brain cured itself when it was no longer exposed to HD's poisonous protein. This mouse then stopped its abnormal movements and was better able to learn again.

Researchers Ai Yamamoto *(left)* and Rene Hen experimented with putting the human HD gene in mice at Dr. Hen's laboratory at Columbia University.

The Hereditary Disease Foundation continued to hold workshops, organize large international meetings, and fund a great deal of HD research. Other groups helped as well.

~ Scientists at Play

Dennis Shea, a retired businessman who had started his own foundation to provide attention to care issues for HD families, built a conference bungalow on his property in Florida. There, the Hereditary Disease Foundation, of which he was a trustee, and Dennis's foundation held joint workshops that were far more relaxed than most scientific meetings. Scientists wandered on the beach dressed in swimsuits and carrying briefcases stuffed with data. Others planned experiments while lounging in the Jacuzzi. The researchers found time to listen to music at nearby hangouts, enjoy beach cookouts, and give parties where having fun appeared to replace most thoughts of serious research. Although the get-togethers seemed laid back, Dennis himself was always in a hurry to find a cure for HD. His two children were at risk of inheriting the disease from their mother.

Partying on Dennis's beach was only one example of HD researchers enjoying life outside the lab. The Hereditary Disease Foundation often gave lively parties during workshops and conferences. When actress Carrie Fisher, best known as Princess Leia in the *Star Wars* films,

> Scientists wandered on the beach dressed in swimsuits and carrying briefcases stuffed with data. Others planned experiments while lounging in the Jacuzzi.

became a trustee of the foundation, she opened her home for the cause and all of Los Angeles showed up with enthusiasm.

In 1998 Alice and Nancy and the Foundation gave Milton Wexler his own party—on the occasion of his 90th birthday and the foundation's 30th anniversary. Trustees Carol Burnett, Julie Andrews, and Sally Kellerman all sang "Happy Birthday." Thirty years had passed since Milton had told his daughters of their mother's illness and their own risk. At this special occasion, Milton, Nancy, and Alice Wexler toasted the accomplishments of both Milton and the foundation.

~ From Bar to Nursing Home

Another happy and noisy party took place in 1999 in the Lake Maracaibo region of Venezuela. This party would inaugurate the Casa Hogar, the clinic and nursing home that replaced what had once been a rowdy bar.

After many years of renovation and fund-raising, this greatly needed facility was finally open and ready for clients. The Red Bull Bar had become the Casa Hogar Corea de Huntington, Amor y Fe. Its name translates as the "Huntington's Chorea Home of Love and

Faith." In the words of Nancy Wexler and Dr. Margot de Young: "Love because that is the first medicine for patients, and faith in God and science to find the cure."

At the inaugural party for the Casa Hogar, guests included the mayor of San Luis, the governor of the province, and the U.S. ambassador to Venezuela. Lively music and plates piled high with food fueled the celebration. The big smiles of Nancy, the ambassador, and Dr. de Young in the party photos tell the story of this happy occasion.

An important guest that day was Dr. Americo Negrette, the scientist who had originally diagnosed Huntington's in the Lake Maracaibo population in 1952. Dr. Negrette, Nancy, and Dr. de Young all received awards for their work. The American ambassador, John Maisto, paid special tribute to Dr. de Young and Nancy: "Through these two professional women—one Venezuelan and one American— we have achieved a marvelous cultural and scientific interchange that will benefit all who are involved in the advancement of science and the treatment of persons with Huntington's."

The Casa Hogar radically transformed the lives of hundreds of people in the Lake Maracaibo region. Where sick villagers had

From a rowdy bar to a haven for Huntington's sufferers. The people who accomplished the miraculous transition of the Casa Hogar include (left to right): San Luis mayor Saadi Bijani, first lady of Zulia State Gladys de Arias Cardenas, Nancy, U.S. ambassador to Venezuela John Malsto, and Dr. Margot de Young cutting the ribbon.

Mayor Bijani presents a medal to Dr. Americo Negrette (left), whose film of the Lake Maracaibo villagers first inspired Nancy to research HD in the region.

Dr. Anne Young, chair of the neurology department of Harvard Medical School, gives "neuros" but here is more friend than physician to a young Venezuelan woman.

once begged and starved in the streets, they could now be fed many meals a day, supplying all the calories they needed to fuel their uncontrollable movements.

Some patients were so sick they lived at the nursing home. Others came during the day for food and medical care. Family members of those too sick to leave home dropped by to pick up food for their relatives.

Today the Casa Hogar continues to offer Huntington's patients clean beds and clean clothes plus a strong sense of dignity and well-being. The home is sparkling clean, and all the staff wear spotless uniforms. The employees are all relatives of HD patients, because Margot and Nancy have a rule that family members get precedence for jobs. Their work at the Casa Hogar helps furnish some of the money they need to care for sick family members. The Home of Love and Faith has proved to be aptly named indeed!

~ One More Celebration

Milton Wexler, then 91, and daughter Nancy enjoy hanging out with members of the Red Hot Chili Peppers band at a 1999 benefit performance.

Huntington's disease supporters closed out the 20th century with another happy event. In 1999 the Red Hot Chili Peppers shook the Los Angeles Palladium with a sold-out benefit performance. The concert raised funds for both the Hereditary Disease Foundation and a musicians' benefit organization. The Chili Peppers blasted the California night with old favorites and new tunes—a red-hot gift to genetic research.

The crowd left the concert hoping that the 21st century would not be very old before they could celebrate the HD cure.

~ Scientific Progress

For the first few years of the 21st century, Nancy Wexler and the HD research team made their annual spring trip to Lake Maracaibo. As usual, before they left for their seven-week stay, they shipped to Venezuela the clothes and other useful items they had been collecting all year: sweatshirts and sweaters to keep the men warm when they fished at night; shorts and T-shirts for everyone on broiling summer days; blankets, towels, and toys for the children; and medicines for all. Once back in the villages, team members made their usual visits to do neuros, draw blood, and renew friendships.

Since the 2002 visit, sadly, the team has been unable to return. After two decades of close ties between the Venezuelan villagers and the American scientists, the U.S. State Department advised against travel there due to an uncertain political situation. "We live in hope that things will change soon," Nancy says. "We're eager to get back and be with our friends again."

The many years of work, however, continue to advance science. The DNA samples and the family pedigree with its complex mapped relationships point the way to more and more helpful information. For example, the HD researchers used to think that the age of onset—that is, the age when HD symptoms first appear—was determined only by the HD gene itself and the number of CAG repeats. Now, using data from 443 Venezuelans, Nancy and her team have found that the age of onset is also influenced by other genes, and by factors in the person's environment. Because the Venezuelan families have an earlier age of onset than do families in the United States or Canada, scientists must nail down what's causing the difference: Is it genetics, environmental factors, or a combination of both?

Scientists are working toward devising drugs that keep people who carry the abnormal HD gene from developing Huntington's

Although she has no children of her own, Nancy feels a tremendous love, kinship, and responsibility for the many thousands of children in families with HD throughout the world.

disease early in life. Such drugs might have to imitate the protective action of certain environmental or genetic factors that may delay onset. As Nancy puts it, "We hope our project will find the factors that offer

Their research is like fitting pieces into a giant jigsaw puzzle; each new discovery helps complete the overall picture.

treatments and cures. We do not merely want to make the disease milder. Our aim is to stop the disease altogether—either by preventing it from appearing or by pushing the age of onset to 140 years, well beyond our life span."

~ Therapies of the Future

Nancy and her team are not the only scientists seeking a cure for Huntington's disease. In 1997 the Hereditary Disease Foundation launched the Cure HD Initiative, a fast track effort to find a cure. Molecular biologist Carl Johnson ran the program and serves today as HDF's executive director for science. Researchers are struggling to pinpoint what makes the disease appear in the first place and what causes it to progress. Funded by the HDF and other groups, they are using animals such as mice, zebra fish, jellyfish, fruit flies, and even a small roundworm.

Scientists have learned much from working with these animals. Their research is like fitting pieces into a giant jigsaw puzzle; each new discovery helps complete the overall picture. Although no one has been able to assemble all the pieces yet, the puzzle is slowly but surely taking shape. And Nancy Wexler has been in the forefront of scientists urging its completion.

~ The Wexlers Today

Nancy's sister, Alice, wrote of the Wexler family and of the search for the HD marker and gene in her book *Mapping Fate: A Memoir of Family, Risk, and Genetic Research*. "It's a little like a family business,"

Alice says of the Wexlers' dedication to the HD cause. "We always talk about it. I find it intellectually exciting. Sometimes working on it is a way of distancing yourself from tragedy and making it bearable."

Both Alice and Nancy are now somewhat beyond the age when Huntington's disease is most likely to be diagnosed, though it's still possible for HD to reveal itself at later ages. Their genetic fate still shadows them. But in addition to themselves, they are committed to finding a cure for everyone in the world as fast as possible. It's a crucial race against time. Nancy takes an upbeat view: "When I walk on a curb and don't fall off, I tell myself I don't have Huntington's!"

Milton Wexler, 97, is legally blind because of another disease, macular degeneration. Yet he still meets with some patients. "If I didn't treat patients," he says, "I would have no function in life." Milton also remains deeply involved as chairman of the board of the Hereditary Disease Foundation. He keeps up with advances in the science of hereditary diseases.

Alice, Milton, and Nancy—a father and two daughters who have dedicated much of their lives to the fight against HD, and are not about to stop.

In addition to Alice and Milton, Nancy considers all the Venezuelans part of her family too. It's hard to believe, but the family with 14 children that Nancy met on her first trip now includes 70 grandchildren and great-grandchildren! With today's knowledge, it is possible to determine that some of the 14 children who had two parents with HD are homozygotes. That is, they have inherited the Huntington's disease gene from both parents. "We were expecting the disease picture to be twice as bad in people who are homozygotes," Nancy says of her last visit to Venezuela. "But they seem to be exactly the same clinically as their brothers and sisters who had inherited HD from only one parent. This is unusual for genetic diseases and was quite unexpected."

Nancy rounds out her report: "We've been trying to investigate other genetic illnesses for which large study populations are needed. Most people think of genetic diseases as some small separate category of illness. Strange as it seems, though, the public is probably more vulnerable to genetic diseases than anything else. Arthritis, heart disease, obesity, and diabetes are all inherited. Conversely, so is longevity." Nancy feels we could all learn a lot about ourselves if we charted our family tree and learned what diseases our relatives had, so we and our doctors would have better information about what to be alert for.

Nancy's passion and mission in life is to find treatments and cures for Huntington's disease as fast as possible. She feels fortunate to have the opportunity to do this important work.

~ Nancy's Gift

Nancy Wexler's search for a cure for Huntington's and other genetic diseases continues today. Many scientists feel that her scientific work plus her role as a "cheerleader" in organizing and encouraging research were largely responsible for the HD gene being found.

As a result of her efforts, people around the world are fighting Huntington's in many ways. Scientists are doing experiments they hope will yield a cure. Support groups around the United States and in dozens of other countries help patients and those at risk share their experiences and their hopes.

Individuals have become involved in unusual ways. For example, Mike and Chris O'Brien, 30-something brothers from New York State, came up with a unique way to contribute to the HD cause: They set out to climb Mount Everest, the world's highest peak, to raise money and awareness for the Hereditary Disease Foundation. Mike and Chris are the fourth and seventh children of a doctor and a nurse. The HD gene killed their mother and a sister and is attacking another sister. Tragically, Mike died in an accident on the mountain on May 1, 2005.

Mike and Chris O'Brien, whose mother and sister had died from HD, are photographed on a climb on Mount Everest designed to raise funds and awareness to benefit the HD cause. Tragically, Mike was killed during the climb, but their message lives on.

For Nancy, science is the most exciting adventure of all. "I love what I'm doing," she says. "Taking part in the struggle against hereditary disease has given my life purpose and direction. I have a chance to participate in groundbreaking research—and, hopefully, the privilege and joy of saving lives."

As she talks, Nancy is preparing to take off for yet another scientific conference. This time it's the annual meeting of the Society for Neuroscience. She's looking forward to seeing old friends again, learning what's new in their research, and going to the parties that are sure to be part of the conference.

Before she leaves, maybe Nancy and Herb can find time for a movie. "Maybe even two in one day!" Or maybe her next trip won't involve work at all. After all, she and Herb are long overdue for a vacation. Maybe this will be the year for Italy.

Nancy sums up her life in science this way: "When we found out about Huntington's in our family, one of the worst aspects for me was the feeling that there was nothing I could do. My father told us there was no cure, but he never said there is no hope."

The Venezuelan families have joined hands across the world with Nancy's family and others to find the cure for Huntington's disease. All these people have made an enormous difference in the advance of medicine—and they show no sign of stopping. That's a pretty impressive gift.

Timeline of Nancy Wexler's Life

1945   Nancy Wexler is born on July 19, the younger sister of Alice, age three.

1961   Nancy's uncle, Jesse Sabin, dies of Huntington's disease (HD).

1962   Paul Sabin dies of HD.

1963   Nancy graduates from high school in Los Angeles, California.

1965   Seymour Sabin dies of HD.

1967   Nancy earns a bachelor's degree in psychology and social relations
       cum laude from Radcliffe College of Harvard University in Cambridge,
       Massachusetts. She spends a Fulbright fellowship year at the University
       of West Indies in Kingston, Jamaica, and the Hampstead Clinic in
       London, England.

1968   Leonore Wexler, Nancy's mother, is diagnosed with HD. Milton Wexler,
       Nancy's father, founds the organization that will become the Hereditary
       Disease Foundation.

1974   Nancy earns a Ph.D. in psychology from the University of Michigan.
       She begins teaching psychology at the New School for Social
       Research in New York City.

1976   Nancy moves to Washington, D.C., to become executive director of
       the Congressional Commission for the Control of Huntington's
       Disease and Its Consequences.

1978   Leonore Wexler dies of HD.

1979   Nancy travels with researcher Tom Chase on her first trip to the Lake
       Maracaibo region of Venezuela to collect blood samples.

1981   The first major research trip to Lake Maracaibo takes place for Nancy
       and others in the United States–Venezuela Collaborative Research
       Project.

1983 Nancy and other scientists report finding a DNA marker genetically linked to the HD gene.

1984 Nancy becomes a professor at Columbia University in New York City.

1986 The Wexler family and HD are featured on an episode of *60 Minutes*.

1992 *Time* magazine runs a featured profile of Nancy.

1993 Nancy and others in the Huntington's Disease Collaborative Group (the Gene Hunters) report finding the gene for HD. Nancy receives an Albert Lasker Public Service Award.

1995 Alice Wexler, Nancy's sister, publishes *Mapping Fate,* the story of the Wexler family and the search for the HD gene.

1996 Nancy and other researchers spend the next nine years using mice and other laboratory animals to advance research on HD.

1997 Bard College awards Nancy the honorary degree of doctor of science. The Cure Huntington's Disease Initiative, a part of the Hereditary Disease Foundation, is launched.

1999 The Casa Hogar, a care facility and home for HD patients, opens in Maracaibo, Venezuela.

2002 The Hereditary Disease Foundation, with Nancy as president, initiates collaborations with investigators to screen drugs for HD treatment.

2004 Nancy and the United States–Venezuela Collaborative Research Project show that age of onset is not solely determined by the HD gene.

2005 Nancy and the Gene Hunter team continue to search for the cure for HD.

About the Author

Adele Glimm's previous biographies of women in science are *Rachel Carson: Protecting Our Earth* and *Elizabeth Blackwell: First Woman Doctor of Modern Times*. Her short stories have appeared in many publications in many countries. Her articles on the craft of writing appear in *The Writer* magazine, and she has taught graduate-level writing courses at Stony Brook University on Long Island. Adele lives in New York City and in Stony Brook, New York, with her husband, Jim.

GLOSSARY

This book is about a neuropsychologist. To figure out the meaning of scientific words, it helps to know a little Greek or Latin. The prefix *neuro* comes from the Greek word *neuron*, meaning "nerve." Neurology is the study of the central nervous system and its diseases. Psychology is the study of the behavior, emotions, and mental state of people. *Psycho* comes from the Greek *psyche*, meaning "mind." A neuropsychologist studies the relationship between disorders of the central nervous system and human behavior.

Here are some other science words you will find in this book. For more information about each word, consult your dictionary.

age of onset: the age of a person when symptoms of a disease first appear

anthropology: the study of humans, especially their physical remains, artifacts, and cultures. A psychiatric anthropologist studies the mental makeup and behavior of entire cultures.

autosome: any chromosome other than a sex chromosome. An autosomal genetic disease is one that both males and females can inherit.

chromosome: a tiny, rod-shaped structure composed of DNA and found in the nucleus of cells of living things. Chromosomes carry the genes that determine heredity. Most humans have 23 pairs of chromosomes.

degenerative: a progressive decline or deterioration in physical and mental abilities

DNA (deoxyribonucleic acid): a molecule in the chromosomes of living cells that carries the genetic code. DNA is a long molecule shaped like a spiral ladder, called a double helix.

dominant: a form of a gene that causes the appearance of a trait regardless of whether an individual inherits one or two copies of the gene

founder: the first member of a family to have a genetic disease that is passed on from generation to generation

gene: the part of the chromosome, consisting of DNA, that influences the inheritance and development of a characteristic passed on from parents to their offspring. From the Greek *genea*, meaning "breed" or "kind."

genetics: a branch of science that deals with the principles of heredity and variations in organisms of the same kind

genome: the complete set of chromosomes with the genes they contain

homozygote: a person who has inherited two copies of either a normal or abnormal gene, one from each parent

Huntington's disease (HD): a degenerative genetic disease of the central nervous system characterized by involuntary physical movements and personality changes. At first, the disease was called Huntington's chorea from the Greek word *choreia,* meaning "dance."

marker: a segment of DNA that can vary between individuals. Looking at patterns of inheritance of markers can help identify the location of genes associated with a particular trait.

nucleotide: any of four chemical compounds that are the basic structural units of DNA. The four nucleotides are adenosine (A), cytosine (C), guanine (G), and thymine (T).

pedigree: a family tree composed of everyone whose ancestry can be traced back to a common ancestor. Pedigree charts help genetic researchers identify which family members passed on particular traits to whom.

protein: any complex combination of amino acids essential to the structure and functioning of all living cells. From the Greek *protos,* meaning "first." One gene makes a protein called "huntingtin." When this protein expands with too many glutamines (an amino acid), Huntington's disease results, causing the destruction of nerve cells in the brain.

psychoanalyst: a person who treats patients with mental, emotional, or behavioral disorders by helping them discover their unconscious fears, feelings, or experiences that may have caused the disorder

sociology: the study of the collective behavior of organized groups of people. From the Latin *socius,* meaning "companion."

FURTHER RESOURCES

Women's Adventures in Science on the Web

Now that you've met Nancy Wexler and learned all about her work, are you wondering what it would be like to be a neuropsychologist? How about a forensic anthropologist, a wildlife biologist, or a robot designer? It's easy to find out. Just visit the *Women's Adventures in Science* Web site at **www.iWASwondering.org.** There you can live your own exciting science adventure. Play games, enjoy comics, and practice being a scientist. While you're having fun, you'll also get to meet amazing women scientists who are changing our world.

BOOKS

Allan, Tony. *Understanding DNA: A Breakthrough in Science.* Chicago: Heinemann Library, 2002. This book examines the history of heredity studies, from Gregor Mendel to James D. Watson and Francis Crick in the 1950s—all of whom contributed to solving the mystery of DNA. Lots of photographs, fact boxes, and timelines illustrate these fascinating stories.

Balkwill, Fran, and Mic Roth. *Have a Nice DNA. Enjoy Your Cells Series.* Cold Spring Harbor, NY: Cold Spring Harbor Laboratory Press, 2002. This book, along with others in the *Enjoy Your Cells Series*, takes you on an entertaining exploration of the amazing hidden world of cells, proteins, and DNA. Great stories plus lively jokes and illustrations make the journey loads of fun.

Silverstein, Alvin, Virginia Silverstein, and Laura Silverstein Nunn. *DNA.* Brookfield, CT: Twenty-First Century Books, 2002. Explore the mysteries of DNA and find out how scientists are working to unravel them. Learn about amino acids, RNA, and the structure of DNA; discover how heredity works; and find out about mutations and genetic disorders. The Human Genome Project, DNA fingerprinting, and genetic engineering also are discussed.

Werlin, Nancy. *Double Helix.* New York: Dial Books, 2004. A suspenseful novel with believable characters, featuring a young hero at risk for Huntington's disease. The subplot, involving genetic engineering, raises important questions.

Yount, Lisa. *Genetics and Genetic Engineering.* New York: Facts on File, 1997. Mendel, DNA, cancer genes, and gene therapy are all discussed in intriguing chapters with useful timelines. One chapter is devoted to Nancy Wexler, the search for the Huntington's gene, and genetic testing.

WEB SITES

Genetic Science Learning Center: http://gslsc.genetics.utah.edu
The University of Utah has a great Web site where you can find out more about how genetic disorders are tracked through families. The interactive Biotechniques Laboratory lets you try actual techniques used by molecular biologists.

Ology, the Gene Scene: http://www.ology.amnh.org/genetics/index.html
The American Museum of Natural History in New York City invites you to go on a genetics journey. Become a DNA detective, extract DNA at home, make a DNA bracelet, and enjoy other fun activities.

Howard Hughes Medical Institute:
http://www.hhmi.org/genetictrail/a100.html#TOP
In the story "Stalking a Lethal Gene," you'll meet Jeff Pinard, a gifted college student with a very serious hereditary disease called cystic fibrosis. Despite major health challenges, Jeff worked as a volunteer in a laboratory, where he researched his own and other genetic mutations by analyzing DNA fragments.

SELECTED BIBLIOGRAPHY

In addition to interviews with Nancy Wexler and her family, friends, and colleagues, the author did extensive reading and research to write this book. Here are some of the sources she consulted.

Angier, Natalie. *The Beauty of the Beastly: New Views on the Nature of Life.* Boston: Houghton Mifflin, 1995.

Bodmer, Walter, and Robin McKie. *The Book of Man: The Human Genome Project and the Quest to Discover Our Genetic Heritage.* New York: Scribner, 1995.

Davies, Kevin. *Cracking the Genome: Inside the Race to Unlock Human DNA.* New York: The Free Press, 2001.

Jones, Steve. *The Language of Genes: Solving the Mysteries of Our Genetic Past, Present, and Future.* New York: Anchor Books, 1995.

Ridley, Matt. *Genome, the Autobiography of a Species in 23 Chapters.* New York: HarperCollins, 1999.

Rothman, Barbara Katz. *Genetic Maps and Human Imaginations: The Limits of Science in Understanding Who We Are.* New York: W. W. Norton, 1998.

Wexler, Alice. *Mapping Fate: A Memoir of Family, Risk, and Genetic Research.* Berkeley: University of California Press, 1995.

INDEX

Illustration Credits:

Except as noted, all photos courtesy Wexler family

Hereditary Disease Foundation (HDF)

National Institute of Neurological Disorders and Strokes (NINDS)

Cover, viii Roy Gumpel; **x** Nicholas Kelsh, courtesy HDF; **1, 2** www.peterginter.com; **5** Steve Uzzell, courtesy HDF; **6** © 1998 PhotoDisc; **9** U.S. National Library of Medicine; **11** (*t*) Carlos and Karen Urrutia, courtesy HDF; (*b*) from *Mapping Fate*, by Alice Wexler, ©. Reprinted by permission of Times Books, a division of Random House Inc. Used by permission of Times Books, a division of Random House Inc.; **12** (*t*) © Douglas Kirkland/Corbis; (*b*) Courtesy Gehry Partners, LLP; **13** Alice Wexler; **22** (*t*) Courtesy Ann Lorimer; **27** © Alain Le Garsmeur/Corbis; **28** (*t*) Radcliffe Archives, Radcliffe Institute, Harvard University; **30** © Daniel Lainé/Corbis; **31** © Bettmann/Corbis; **35** Library of Congress; **40** New York Brain Bank at Columbia University; **42** www.peterginter.com; **43** Courtesy Nancy Wexler and HDF; **44** © Morton Beebe/Corbis; **45, 49** Courtesy Nancy Wexler and HDF; **54** Steve Uzzell, courtesy HDF; **55** Alice Wexler; **56** Courtesy NINDS; **57** www.peterginter.com; **59, 61** Steve Uzzell, courtesy HDF; **62** Alice Wexler; **63** www.peterginter.com; **64** Steve Uzzell, courtesy HDF; **66** www.peterginter.com; **67, 68** Steve Uzzell, courtesy HDF; **70** Roy Gumpel; **71** Steve Uzzell, courtesy HDF; **72, 73** Courtesy Nancy Wexler and HDF; **74** Elaine Attias, HDF Trustee; **77** Courtesy Nancy Wexler and HDF; **78** Roy Gumpel; **82** Gustavo Rey, courtesy HDF; **83** © 1998 PhotoDisc; **84, 86** Courtesy Nancy Wexler and HDF; **87** Courtesy Ann Lorimer; **89** Courtesy Nancy Wexler and HDF; **90** Julie Porter, courtesy HDF; **92** Steve Uzzell, courtesy HDF; **94** Courtesy Nancy Wexler and HDF; **95** © 1998 PhotoDisc; **97** Courtesy *P&S Journal*, Columbia University College of Physicians and Surgeons, Winter 2003, Vol. 23, No. 1; **99-101** Courtesy Nancy Wexler and HDF; **103** Michael Collins; **104** © Chris Carroll; **105** (*t*) Courtesy O'Brien family

Maps: © 2004 Map Resources

Illustrations: Max-Karl Winkler

The border image used throughout the book shows neurodegeneration of a brain caused by Huntington's disease.

JHP Executive Editor: Stephen Mautner

Series Managing Editor: Terrell D. Smith

Designer: Francesca Moghari

Illustration research: Joan Mathys

Special contributors: Meredith DeSousa, Sally Groom, Mary Kalamaras, Emily Kohn, Dorothy Lewis, April Luehmann, John Quackenbush, and Anita Schwartz

Graphic design assistance: Michael Dudzik and Anne Rogers